IT'S HERE!

PRENTICE HALL
SCIENCE

FINALLY, THE PERFECT FIT.

NOW YOU CAN CHOOSE THE PERFECT FIT FOR ALL YOUR CURRICULUM NEEDS.

The new Prentice Hall Science program consists of 19 hardcover books, each of which covers a particular area of science. All of the sciences are represented in the program so you can choose the perfect fit to *your* particular curriculum needs.

The flexibility of this program will allow you to teach those topics you want to teach, and to teach them *in-depth*. Virtually any approach to science—general, integrated, coordinated, thematic, etc.—is possible with Prentice Hall Science.

Above all, the program is designed to make your teaching experience easier and more fun.

ELECTRICITY AND MAGNETISM

Ch. 1. Electric Charges and Currents
Ch. 2. Magnetism
Ch. 3. Electromagnetism
Ch. 4. Electronics and Computers

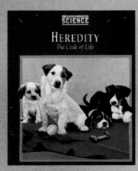

HEREDITY: THE CODE OF LIFE

Ch. 1. What is Genetics?
Ch. 2. How Chromosomes Work
Ch. 3. Human Genetics
Ch. 4. Applied Genetics

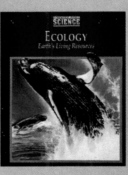

ECOLOGY: EARTH'S LIVING RESOURCES

Ch. 1. Interactions Among Living Things
Ch. 2. Cycles in Nature
Ch. 3. Exploring Earth's Biomes
Ch. 4. Wildlife Conservation

PARADE OF LIFE: MONERANS, PROTISTS, FUNGI, AND PLANTS

Ch. 1. Classification of Living Things
Ch. 2. Viruses and Monerans
Ch. 3. Protists
Ch. 4. Fungi
Ch. 5. Plants Without Seeds
Ch. 6. Plants With Seeds

EXPLORING THE UNIVERSE

Ch. 1. Stars and Galaxies
Ch. 2. The Solar System
Ch. 3. Earth and Its Moon

EVOLUTION: CHANGE OVER TIME

Ch. 1. Earth's History in Fossils
Ch. 2. Changes in Living Things Over Time
Ch. 3. The Path to Modern Humans

EXPLORING EARTH'S WEATHER

Ch. 1. What Is Weather?
Ch. 2. What Is Climate?
Ch. 3. Climate in the United States

THE NATURE OF SCIENCE

Ch. 1. What is Science?
Ch. 2. Measurement and the Sciences
Ch. 3. Tools and the Sciences

ECOLOGY: EARTH'S NATURAL RESOURCES

Ch. 1. Energy Resources
Ch. 2. Earth's Nonliving Resources
Ch. 3. Pollution
Ch. 4. Conserving Earth's Resources

MOTION, FORCES, AND ENERGY

Ch. 1. What Is Motion?
Ch. 2. The Nature of Forces
Ch. 3. Forces in Fluids
Ch. 4. Work, Power, and Simple Machines
Ch. 5. Energy: Forms and Changes

PARADE OF LIFE: ANIMALS

Ch. 1. Sponges, Cnidarians, Worms, and Mollusks
Ch. 2. Arthropods and Echinoderms
Ch. 3. Fish and Amphibians
Ch. 4. Reptiles and Birds
Ch. 5. Mammals

CELLS: BUILDING BLOCKS OF LIFE

Ch. 1. The Nature of LIfe
Ch. 2. Cell Structure and Function
Ch. 3. Cell Processes
Ch. 4. Cell Energy

DYNAMIC EARTH

Ch. 1. Movement of the Earth's Crust
Ch. 2. Earthquakes and Volcanoes
Ch. 3. Plate Tectonics
Ch. 4. Rocks and Minerals
Ch. 5. Weathering and Soil Formation
Ch. 6. Erosion and Deposition

MATTER: BUILDING BLOCK OF THE UNIVERSE

Ch. 1. General Properties of Matter
Ch. 2. Physical and Chemical Changes
Ch. 3. Mixtures, Elements, and Compounds
Ch. 4. Atoms: Building Blocks of Matter
Ch. 5. Classification of Elements: The Periodic Table

CHEMISTRY OF MATTER

Ch. 1. Atoms and Bonding
Ch. 2. Chemical Reactions
Ch. 3. Families of Chemical Compounds
Ch. 4. Chemical Technology
Ch. 5. Radioactive Elements

HUMAN BIOLOGY AND HEALTH

Ch. 1. The Human Body
Ch. 2. Skeletal and Muscular Systems
Ch. 3. Digestive System
Ch. 4. Circulatory System
Ch. 5. Respiratory and Excretory Systems
Ch. 6. Nervous and Endocrine Systems
Ch. 7. Reproduction and Development
Ch. 8. Immune System
Ch. 9. Alcohol, Tobacco, and Drugs

EXPLORING PLANET EARTH

Ch. 1. Earth's Atmosphere
Ch. 2. Earth's Oceans
Ch. 3. Earth's Fresh Water
Ch. 4. Earth's Landmasses
Ch. 5. Earth's Interior

HEAT ENERGY

Ch. 1. What Is Heat?
Ch. 2. Uses of Heat

SOUND AND LIGHT

Ch. 1. Characteristics of Waves
Ch. 2. Sound and Its Uses
Ch. 3. Light and the Electro-magnetic Spectrum
Ch. 4. Light and Its Uses

A COMPLETELY INTEGRATED LEARNING SYSTEM...

The Prentice Hall Science program is an *integrated* learning system with a variety of print materials and multimedia components. All are designed to meet the needs of diverse learning styles and your technology needs.

THE STUDENT BOOK

Each book is a model of **excellent writing and dynamic visuals**—designed to be exciting and motivating to the student *and* the teacher, with relevant examples integrated throughout, and more opportunities for many different activities which apply to everyday life.

Problem-solving activities emphasize the thinking process, so problems may be more open-ended.

"Discovery Activities" throughout the book foster active learning.

Different sciences, and other disciplines, are integrated throughout the text and reinforced in the "Connections" features (the connections between computers and viruses is one example).

TEACHER'S RESOURCE PACKAGE

In addition to the student book, the complete teaching package contains:

ANNOTATED TEACHER'S EDITION

Designed to provide **"teacher-friendly"** support regardless of instructional approach:

■ **Help is readily available** if you choose to teach thematically, to integrate the sciences, and/or to integrate the sciences with other curriculum areas.

■ **Activity-based learning** is easy to implement through the use of Discovery Strategies, Activity Suggestions, and Teacher Demonstrations.

■ Integration of all components is part of the teaching strategies.

■ For instant accessibility, all of the teaching suggestions are wrapped around the student pages to which they refer.

ACTIVITY BOOK

Includes a **discovery activity for each chapter**, plus other activities including problem-solving and cooperative-learning activities.

THE REVIEW AND REINFORCEMENT GUIDE

Addresses **students' different learning styles** in a clear and comprehensive format:

■ Highly visual for visual learners.

TEACHER'S RESOURCE PACKAGE

FOR THE PERFECT FIT TO YOUR TEACHING NEEDS.

■ Can be used in conjunction with the program's audiotapes for auditory and language learners.

■ More than a study guide, it's a guide to comprehension, with activities, key concepts, and vocabulary.

ENGLISH AND SPANISH AUDIOTAPES
Correlate with the Review and Reinforcement Guide to aid auditory learners.

LABORATORY MANUAL ANNOTATED TEACHER'S EDITION
Offers **at least one additional hands-on opportunity per chapter** with

answers and teaching suggestions on lab preparation and safety.

TEST BOOK
Contains **traditional and up-to-the-minute strategies for student assessment.** Choose from performance-based tests in addition to traditional chapter tests and computer test bank questions.

STUDENT LABORATORY MANUAL
Each of the 19 books also comes with its own Student Lab Manual.

ALSO INCLUDED IN THE INTEGRATED LEARNING SYSTEM:

■ Teacher's Desk Reference

■ English Guide for Language Learners

■ Spanish Guide for Language Learners

■ Product Testing Activities

■ Transparencies

■ Computer Test Bank (IBM, Apple, or MAC)

■ VHS Videos

■ Videodiscs

■ Interactive Videodiscs (Level III)

■ Interactive Videodiscs/ CD ROM

■ Courseware

All components are integrated in the teaching strategies in the Annotated Teacher's Edition, where they directly relate to the science content.

THE PRENTICE HALL SCIENCE
INTEGRATED LEARNING SYSTEM

The following components are integrated in the teaching strategies for
CHEMISTRY OF MATTER.

- **Spanish Audiotape English Audiotape**
- **Activity Book**
- **Review and Reinforcement Guide**
- **Test Book**—including Performance-Based Tests
- **Laboratory Manual, Annotated Teacher's Edition**
- **Product-Testing Activities:**
  Testing Antacids
  Testing Cereals
  Testing Nail Enamels

- **Laboratory Manual**
- **English Guide for Language Learners**
- **Spanish Guide for Language Learners**
- **Transparencies:**
  Structure of the Atom
  Energy Levels
  Ionic Bonding
  Periodic Table
  Covalent Bonding
  Molecules
  Acids, Bases, and pH Values

- **Videos/Videodiscs:**
  Chemical Bonding and Atomic Structure
  Elements, Compounds, and Mixtures
  Periodic Table and Periodicity
  Acids, Bases, and Salts
- **Interactive Videodisc:**
  ScienceVision: Chemical Pursuits

INTEGRATING OTHER SCIENCES

Many of the other 18 Prentice Hall Science books can be integrated into **CHEMISTRY OF MATTER.** The books you will find suggested most often in the Annotated Teacher's Edition are MATTER: BUILDING BLOCK OF THE UNIVERSE; CELLS: BUILDING BLOCKS OF LIFE; ELECTRICITY AND MAGNETISM; DYNAMIC EARTH; HEAT ENERGY; MOTION, FORCES, AND ENERGY; HUMAN BIOLOGY AND HEALTH; ECOLOGY: EARTH'S NATURAL RESOURCES; EXPLORING PLANET EARTH; THE NATURE OF SCIENCE; EXPLORING THE UNIVERSE; PARADE OF LIFE: ANIMALS; EVOLUTION: CHANGE OVER TIME; PARADE OF LIFE: MONERANS, PROTISTS, FUNGI, AND PLANTS; and HEREDITY: THE CODE OF LIFE.

INTEGRATING THEMES

Many themes can be integrated into **CHEMISTRY OF MATTER.**
Following are the ones most commonly suggested in the Annotated Teacher's Edition: ENERGY, PATTERNS OF CHANGE, SCALE AND STRUCTURE, SYSTEMS AND INTERACTIONS, and STABILITY.

For more detailed information on teaching thematically and integrating the sciences, see the Teacher's Desk Reference and teaching strategies throughout the Annotated Teacher's Edition.

For more information, call 1-800-848-9500 or write:

P R E N T I C E   H A L L

Simon & Schuster Education Group
113 Sylvan Avenue Route 9W
Englewood Cliffs, New Jersey 07632
Simon & Schuster A Paramount Communications Company

Annotated Teacher's Edition

Prentice Hall Science
Chemistry of Matter

Anthea Maton
Former NSTA National Coordinator
Project Scope, Sequence,
    Coordination
Washington, DC

Jean Hopkins
Science Instructor and Department
    Chairperson
John H. Wood Middle School
San Antonio, Texas

Susan Johnson
Professor of Biology
Ball State University
Muncie, Indiana

David LaHart
Senior Instructor
Florida Solar Energy Center
Cape Canaveral, Florida

Charles William McLaughlin
Science Instructor and Department
    Chairperson
Central High School
St. Joseph, Missouri

Maryanna Quon Warner
Science Instructor
Del Dios Middle School
Escondido, California

Jill D. Wright
Professor of Science Education
Director of International Field
    Programs
University of Pittsburgh
Pittsburgh, Pennsylvania

Prentice Hall
A Division of Simon & Schuster
Englewood Cliffs, New Jersey

ISBN 0-13-400672-0

   3 4 5 6 7 8 9 10   97 96 95 94

Contents of Annotated Teacher's Edition

To the Teacher

Welcome to the *Prentice Hall Science* program. *Prentice Hall Science* has been designed as a complete program for use with middle school or junior high school science students. The program covers all relevant areas of science and has been developed with the flexibility to meet virtually all your curriculum needs. In addition, the program has been designed to better enable you—the classroom teacher—to integrate various disciplines of science into your daily lessons, as well as to enhance the thematic teaching of science.

The *Prentice Hall Science* program consists of nineteen books, each of which covers a particular topic area. The nineteen books in the *Prentice Hall Science* program are

The Nature of Science
Parade of Life: Monerans, Protists, Fungi, and Plants
Parade of Life: Animals
Cells: Building Blocks of Life
Heredity: The Code of Life
Evolution: Change Over Time

Ecology: Earth's Living Resources
Human Biology and Health
Exploring Planet Earth
Dynamic Earth
Exploring Earth's Weather
Ecology: Earth's Natural Resources
Exploring the Universe
Matter: Building Block of the Universe
Chemistry of Matter
Electricity and Magnetism
Heat Energy
Sound and Light
Motion, Forces, and Energy

Each of the student editions listed above also comes with a complete set of teaching materials and student ancillary materials. Furthermore, videos, interactive videos and science courseware are available for the *Prentice Hall Science* program. This combination of student texts and ancillaries, teacher materials, and multimedia products makes up your complete *Prentice Hall Science* Learning System.

About the Teacher's Desk Reference

The *Teacher's Desk Reference* provides you, the teacher, with an insight into the workings of the *Prentice Hall Science* program. The *Teacher's Desk Reference* accomplishes this task by including all the standard information you need to know about *Prentice Hall Science*.

The *Teacher's Desk Reference* presents an overview of the program, including a full description of each ancillary available in the program. It gives a brief summary of each of the student textbooks available in the *Prentice Hall Science* Learning System. The *Teacher's Desk Reference* also demonstrates how the seven science themes incorporated into *Prentice Hall Science* are woven throughout the entire program.

In addition, the *Teacher's Desk Reference* presents a detailed discussion of the features of the Student

Edition and the features of the Annotated Teacher's Edition, as well as an overview section that summarizes issues in science education and offers a message about teaching special students. Selected instructional essays in the *Teacher's Desk Reference* include English as a Second Language (ESL), Multicultural Teaching, Cooperative-Learning Strategies, and Integrated Science Teaching, in addition to other relevant topics. Further, a discussion of the Multimedia components that are part of *Prentice Hall Science*, as well as how they can be integrated with the textbooks, is included in the *Teacher's Desk Reference*.

The *Teacher's Desk Reference* also contains in blackline master form a booklet on Teaching Graphing Skills, which may be reproduced for student use.

Integrating the Sciences

The *Prentice Hall Science* Learning System has been designed to allow you to teach science from an integrated point of view. Great care has been taken to integrate other science disciplines, where appropriate, into the chapter content and visuals. In addition, the integration of other disciplines such as social studies and literature has been incorporated into each textbook.

On the reduced student pages throughout your Annotated Teacher's Edition you will find numbers within blue bullets beside selected passages and visuals. An Annotation Key in the wraparound margins indicates the particular branch of science or other discipline that has been integrated into the student text. In addition, where appropriate, the name of the textbook and the chapter number in which the particular topic is discussed in greater detail is provided. This enables you to further integrate a particular science topic by using the complete *Prentice Hall Science* Learning System.

Thematic Overview

When teaching any science topic, you may want to focus your lessons around the underlying themes that pertain to all areas of science. These underlying themes are the framework from which all science can be constructed and taught. The seven underlying themes incorporated into *Prentice Hall Science* are

Energy
Evolution
Patterns of Change
Scale and Structure
Systems and Interactions
Unity and Diversity
Stability

A detailed discussion of each of these themes and how they are incorporated into the *Prentice Hall Science* program are included in your *Teacher's Desk Reference*. In addition, the *Teacher's Desk Reference* includes thematic matrices for the *Prentice Hall Science* program.

A thematic matrix for each chapter in this textbook follows. Each thematic matrix is designed with the list of themes along the left-hand column and in the right-hand column a big idea, or overarching concept statement, as to how that particular theme is taught in the chapter.

The primary themes in this textbook are Energy, Patterns of Change, Scale and Structure, Systems and Interactions, and Stability. Primary themes throughout *Prentice Hall Science* are denoted by an asterisk.

CHAPTER 1

Atoms and Bonding

***ENERGY**	• Energy is always involved in chemical bonding. • Ionization, or the process of removing electrons from an atom and forming an ion, requires energy (known as ionization energy). When an electron is gained, energy is released.
EVOLUTION	
***PATTERNS OF CHANGE**	• The combining of atoms of elements to form new substances is called chemical bonding. • Ions of opposite charge strongly attract each other.
***SCALE AND STRUCTURE**	• The atom consists of a nucleus (or center), in which are found protons and neutrons, and surrounding energy levels, in which are found electrons. • A crystal lattice is made of huge numbers of ions. • Network solids contain large numbers of atoms covalently bonded to one another.
***SYSTEMS AND INTERACTIONS**	• Ionic bonding involves the transfer of electrons (one atom gains; the other loses) and the formation of ions and crystal lattices. • Covalent bonds involve the sharing of electrons and the formation of molecules. • Metallic bonding involves a sea of mobile electrons.
UNITY AND DIVERSITY	• Atoms will bond in order to achieve stability. The bond type can be ionic, covalent, or metallic. • Regardless of type, all bonds involve electrons—usually the valence electrons.
***STABILITY**	• In order to achieve stability, an atom will either gain or lose electrons and therefore attain a complete outermost energy level. • The sum of the oxidation numbers of the atoms in a compound must be zero.

CHAPTER 2

Chemical Reactions

*ENERGY	• The energy of the reactants changes during a chemical reaction. • A chemical reaction can give off energy (exothermic) or absorb energy (endothermic). • The reactants must be able to reach the activation energy before the products can be formed.
EVOLUTION	
*PATTERNS OF CHANGE	• A chemical reaction changes original substances into new substances with different chemical and physical properties, including a different amount of energy.
*SCALE AND STRUCTURE	• The ability for a substance to undergo a chemical reaction depends on the number of valence electrons in its atoms. • Collisions among the particles of the reactants determine the rate of a chemical reaction.
*SYSTEMS AND INTERACTIONS	• Increasing the temperature, concentration, or surface area of the reactants usually increases reaction rate. • A catalyst can also increase reaction rate by decreasing the required activation energy.
UNITY AND DIVERSITY	• All chemical reactions share certain characteristics, such as changing reactants into products. However, there are different types of chemical reactions: synthesis reactions, decomposition reactions, single-replacement reactions, and double-replacement reactions.
*STABILITY	• Mass is always conserved during a chemical reaction.

CHAPTER 3

Families of Chemical Compounds

*ENERGY	• All chemical reactions involved in the formation of acids, bases, salts, and organic compounds are either endothermic or exothermic.
EVOLUTION	
*PATTERNS OF CHANGE	• The process in which an acid chemically combines with a base is called neutralization. • The products of a neutralization reaction are a salt and water. The pH of the products is 7. • Each alkane, alkene, and alkyne differs from its preceding series member by CH_2.
*SCALE AND STRUCTURE	• Carbon's ability to combine with itself and with other elements in a variety of ways explains why more than 2 million organic compounds exist.
*SYSTEMS AND INTERACTIONS	• The rate of solution depends on temperature, surface area, and agitation. • The solubility of a solute is affected by temperature and pressure. • When acids react chemically with bases, they form salts and water. • Substituted hydrocarbons are formed when one or more hydrogen atoms in a hydrocarbon chain or ring is replaced by a different atom or group of atoms.
UNITY AND DIVERSITY	• All solutions contain a solute dissolved in a solvent. • Solutions can be unsaturated, saturated, or supersaturated, depending on how much solute is dissolved in a given amount of solute at a certain temperature. • Acids are proton donors; bases are proton acceptors. • All organic compounds contain carbon. There are many different groups of organic compounds.
*STABILITY	• A supercooled solution is supersaturated and is highly unstable. Saturated hydrocarbons (alkanes) are more stable than unsaturated hydrocarbons (alkenes and alkynes) and thus tend to be less reactive.

CHAPTER 4

Petrochemical Technology

*ENERGY	• Petroleum is particularly important because a tremendous amount of energy is released when petroleum is burned.
EVOLUTION	• Petroleum is believed to have formed from the remains of dead animals and plants buried in the oceans and subjected to tremendous heat and pressure over millions of years.
*PATTERNS OF CHANGE	• Raw materials from petroleum are converted into chemicals. The chemicals are then used to make various products called petrochemical products.
*SCALE AND STRUCTURE	• Polymers are molecules made of chains of smaller molecular units called monomers.
*SYSTEMS AND INTERACTIONS	• Petroleum can be divided into its various parts, or fractions, in a process known as fractional distillation.
UNITY AND DIVERSITY	• The separation of petroleum into fractions is based on the fact that the fractions have different boiling points. • Some polymers are found naturally; others are produced synthetically from petrochemicals.
*STABILITY	• Petroleum is a nonrenewable resource. Its supply is limited, and once it is diminished, no more will be produced.

CHAPTER 5

Radioactive Elements

*ENERGY	• Binding energy provided by the nuclear strong force holds the nucleus together. If binding energy is low, a nucleus will be radioactive. • Tremendous amounts of energy are released during fission and fusion.
EVOLUTION	• The age of fossils can be determined using the half-lives of certain radioactive elements.
*PATTERNS OF CHANGE	• If an unstable nucleus can become stable by changing, it will undergo a nuclear reaction.
*SCALE AND STRUCTURE	• One individual fission reaction releases a relatively small amount of energy. The neutrons emitted during the reaction, however, cause other nuclei to undergo fission in a chain reaction. The total amount of energy released is tremendous.
*SYSTEMS AND INTERACTIONS	• Bombarding the nucleus with subatomic particles can result in transmutation.
UNITY AND DIVERSITY	• There are four types of nuclear reactions. • Radioactive decay and artificial transmutation involve the emission of fragments of the nucleus. Fission involves splitting a nucleus into two. Fusion involves combining two nuclei to form one.
*STABILITY	• An element that undergoes radioactive decay will continue to decay until a stable nucleus is achieved. The various elements produced in the process constitute a decay series.

Comprehensive List of Laboratory Materials

Item	Quantities per Group	Chapter
Acid, mild (lime or lemon juice or vinegar)	150 mL	4
Acid solutions (H_2SO_4, HCl, HNO_3)	5 mL	3
Basic solutions (KOH, NaOH, Ca(OH_2)	5 mL	3
Beakers		
100-mL	2	1
250-mL	3	2
medium	1	3
Bleach, liquid	200 mL	4
Bowl, large	1	5
Bunsen burner	1	1
Cloth samples, 15 cm × 15 cm		
polyester, nylon, acetate	1 of each	4
wool, cotton, liner	1 of each	4
Connecting wires	3	1
Cups, Styrofoam	12	4
Dry cell	1	1
Evaporating dish	1	3
Food coloring	1 small bottle	5
Gloves, rubber	1 pair	4
Graduated cylinders	2	2
Light bulb	1	1
Light bulb socket	1	1
Litmus paper, red and blue	small box	3
Medicine dropper	1	3, 4, 5
Metric ruler	1	4
Oil, vegetable	50 mL	1
	6 drops	4
Paper towel	1	4
Pen, marking	1	4
Pencil, glass-marking	1	1
Phenolphthalein	few drops	3
Salt	small sample	1
Scissors	1	4
Stirring rod	1	1, 2, 3
Stopwatch (or watch with sweep second hand)	1	2
Sugar	small sample	1
Sugar cubes	250	5
Test tubes, medium	4	1
	6	3
Test-tube rack	1	3
Test-tube tongs	1	1
Timer	1	1
Water, distilled	200 mL	1, 2

CHEMISTRY OF MATTER

Anthea Maton
Former NSTA National Coordinator
Project Scope, Sequence, Coordination
Washington, DC

Jean Hopkins
Science Instructor and Department Chairperson
John H. Wood Middle School
San Antonio, Texas

Susan Johnson
Professor of Biology
Ball State University
Muncie, Indiana

David LaHart
Senior Instructor
Florida Solar Energy Center
Cape Canaveral, Florida

Charles William McLaughlin
Science Instructor and Department Chairperson
Central High School
St. Joseph, Missouri

Maryanna Quon Warner
Science Instructor
Del Dios Middle School
Escondido, California

Jill D. Wright
Professor of Science Education
Director of International Field Programs
University of Pittsburgh
Pittsburgh, Pennsylvania

Prentice Hall
Englewood Cliffs, New Jersey
Needham, Massachusetts

Prentice Hall Science
Chemistry of Matter

Student Text and Annotated Teacher's Edition
Laboratory Manual
Teacher's Resource Package
Teacher's Desk Reference
Computer Test Bank
Teaching Transparencies
Product Testing Activities
Computer Courseware
Video and Interactive Video

The illustration on the cover, rendered by David Schleinkofer, shows a display of fireworks made possible by interactions of matter.

Credits begin on page 174.

SECOND EDITION

ISBN 0-13-400656-9

3 4 5 6 7 8 9 10   97 96 95 94

Prentice Hall
A Division of Simon & Schuster
Englewood Cliffs, New Jersey 07632

STAFF CREDITS

Editorial:	Harry Bakalian, Pamela E. Hirschfeld, Maureen Grassi, Robert P. Letendre, Elisa Mui Eiger, Lorraine Smith-Phelan, Christine A. Caputo
Design:	AnnMarie Roselli, Carmela Pereira, Susan Walrath, Leslie Osher, Art Soares
Production:	Suse F. Bell, Joan McCulley, Elizabeth Torjussen, Christina Burghard
Photo Research:	Libby Forsyth, Emily Rose, Martha Conway
Publishing Technology:	Andrew Grey Bommarito, Deborah Jones, Monduane Harris, Michael Colucci, Gregory Myers, Cleasta Wilburn
Marketing:	Andrew Socha, Victoria Willows
Pre-Press Production:	Laura Sanderson, Kathryn Dix, Denise Herckenrath
Manufacturing:	Rhett Conklin, Gertrude Szyferblatt

Consultants

Kathy French	National Science Consultant
Jeannie Dennard	National Science Consultant
Brenda Underwood	National Science Consultant
Janelle Conarton	National Science Consultant

CONTENTS

CHEMISTRY OF MATTER

SCIENCE GAZETTES

Activity Bank/Reference Section

Features

CONCEPT MAPPING

Throughout your study of science, you will learn a variety of terms, facts, figures, and concepts. Each new topic you encounter will provide its own collection of words and ideas—which, at times, you may think seem endless. But each of the ideas within a particular topic is related in some way to the others. No concept in science is isolated. Thus it will help you to understand the topic if you see the whole picture; that is, the interconnectedness of all the individual terms and ideas. This is a much more effective and satisfying way of learning than memorizing separate facts.

Actually, this should be a rather familiar process for you. Although you may not think about it in this way, you analyze many of the elements in your daily life by looking for relationships or connections. For example, when you look at a collection of flowers, you may divide them into groups: roses, carnations, and daisies. You may then associate colors with these flowers: red, pink, and white. The general topic is flowers. The subtopic is types of flowers. And the colors are specific terms that describe flowers. A topic makes more sense and is more easily understood if you understand how it is broken down into individual ideas and how these ideas are related to one another and to the entire topic.

It is often helpful to organize information visually so that you can see how it all fits together. One technique for describing related ideas is called a **concept map**. In a concept map, an idea is represented by a word or phrase enclosed in a box. There are several ideas in any concept map. A connection between two ideas is made with a line. A word or two that describes the connection is written on or near the line. The general topic is located at the top of the map. That topic is then broken down into subtopics, or more specific ideas, by branching lines. The most specific topics are located at the bottom of the map.

To construct a concept map, first identify the important ideas or key terms in the chapter or section. Do not try to include too much information. Use your judgment as to what is really important. Write the general topic at the top of your map. Let's use an example to help illustrate this process. Suppose you decide that the key terms in a section you are reading are School, Living Things, Language Arts, Subtraction, Grammar, Mathematics, Experiments, Papers, Science, Addition, Novels. The general topic is School. Write and enclose this word in a box at the top of your map.

SCHOOL

Now choose the subtopics—Language Arts, Science, Mathematics. Figure out how they are related to the topic. Add these words to your map. Continue this procedure until you have included all the important ideas and terms. Then use lines to make the appropriate connections between ideas and terms. Don't forget to write a word or two on or near the connecting line to describe the nature of the connection.

Do not be concerned if you have to redraw your map (perhaps several times!) before you show all the important connections clearly. If, for example, you write papers for Science as well as for Language Arts, you may want to place these two subjects next to each other so that the lines do not overlap.

One more thing you should know about concept mapping: Concepts can be correctly mapped in many different ways. In fact, it is unlikely that any two people will draw identical concept maps for a complex topic. Thus there is no one correct concept map for any topic! Even

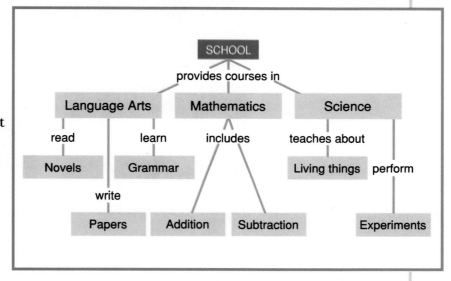

though your concept map may not match those of your classmates, it will be correct as long as it shows the most important concepts and the clear relationships among them. Your concept map will also be correct if it has meaning to you and if it helps you understand the material you are reading. A concept map should be so clear that if some of the terms are erased, the missing terms could easily be filled in by following the logic of the concept map.

Chemistry of Matter

In this textbook students are introduced to the chemical properties of matter. Students first explore chemical bonding. They learn about ionic, covalent, and metallic bonds and about how to predict bond type. They are introduced to chemical reactions and chemical equations and to the energy associated with the chemical reactions. The topic of reaction rate is also covered, and students apply the collision theory to explain factors that affect rate.

Next, students learn about solutions and their properties. They define acids, bases, and salts. They then explore the chemistry of carbon compounds, identifying them as the most numerous compounds. Next, they examine petrochemical technology and describe the process of polymerization. Finally, they study the properties of radioactive elements. They study nuclear reactions including transmutation, fission, and fusion. They also learn about the uses and dangers of radioactivity.

TEXT OBJECTIVES

1. Explain chemical bonding and contrast ionic, covalent, and metallic bonding.
2. Describe and explain the basis of chemical reactions and interpret and balance equations.
3. Contrast exothermic and endothermic reactions and explain the factors that affect reaction rates.
4. Define solution and classify solutions.
5. Describe the properties of acids and bases and the properties of salts.
6. Classify different carbon compounds and draw structural formulas for them.
7. Identify petrochemical products and explain the process of polymerization.
8. Contrast different types of nuclear reactions and describe the detection, measurement, and uses of radioactivity.

CHEMISTRY OF MATTER

The fun of blowing bubbles is made possible by chemistry.

You have been given an assignment to perform the following steps, but you have not been told what the results of the procedure will be. What will you have made after completing the steps? Read them and see if you can figure it out. First, treat a fat called palmitin with an alkali such as sodium hydroxide in a process called saponification. The fat will break down to produce the substances sodium palmitate and glycerin. Discard the glycerin. Then add the sodium palmitate to a wetting compound to form a solution. Now dip a thin ring, preferably one with a handle, into the liquid. Finally, apply a gentle stream of air to the film on the ring.

Well, have you figured it out? Do you know what you have done? This rather complicated

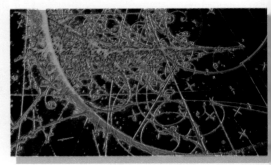

The activities of the tiny particles that make up matter can be traced in patterns such as this one.

INTRODUCING CHEMISTRY OF MATTER

USING THE TEXTBOOK

Begin your teaching of the textbook by having students examine the textbook-opening photographs and captions. Before students read the textbook introduction, ask them the following questions:
• **How does blowing bubbles involve chemistry?** (Soap is a chemical compound.)

• **How is chemistry involved in digestion?** (A number of chemical processes enable a person to eat, digest, and process foods to sustain life through chemical processes.)

Have students read the textbook introduction. Based on the pictures, captions, and introduction, what do they think saponification is? (Many students will surmise that it is the making of soap.) Have students use a dictionary to define *saponification*. (The conversion of fat into soap.)

CHAPTERS

procedure has a fairly simple explanation: When you saponified the palmitin, you made soap. Then you added it to water to make a solution. And finally, you blew bubbles!

Chemical reactions such as saponification may not be familiar to you. Yet chemical reactions are occurring all around you and even in your body at this very moment. As you read this textbook, you will learn about the interactions of matter that can occur in a test tube, in nature, and even inside you!

Some of the most important chemical activities occur during the simple act of eating.

Discovery Activity

Chemical Mysteries

1. Dip a toothpick into a small dish of milk. Use the toothpick as a pen to write a message on a sheet of white paper.

2. Let the milk dry. Observe what happens to your message as it dries.

3. Hold the paper close to a light bulb that is lit.
   ■ How does your message change before and after you place it next to the light bulb? What can you say about the relationship between the milk and the light bulb?

4. Fill one container about half full with very cold water and another container about half full with hot water.

5. Place four or five drops of food coloring into each container at exactly the same time.
   ■ Compare the rates at which the colors spread. Explain your observations.

O ■ 9

CHAPTER DESCRIPTIONS

1 Atoms and Bonding In Chapter 1, the basis of chemical bonding is explained. Ionic, covalent, and metallic bonding are compared and contrasted. Methods for predicting bond types are also explained.

2 Chemical Reactions Chapter 2 focuses on the nature of chemical changes. Students learn to balance chemical equations and find out about the differences between endothermic and exothermic reactions. The chapter includes a discussion of reaction rate, activation energy, and collision theory.

3 Families of Chemical Compounds Chapter 3 deals with the nature of solutions and the factors that affect the rate of solution and solubility. Acid-base chemistry and pH values are explored, as is the chemistry of organic compounds. Saturated, unsaturated, and supersaturated hydrocarbons as well as other families of organic compounds are described.

4 Petrochemical Technology In Chapter 4, petrochemical processes are explored. Students develop an understanding of the processes and products that have made petroleum one of the most important resources in the twentieth century.

5 Radioactive Elements Chapter 5 deals with the changes in the nuclei of atoms. Radioactivity and its discovery, measurement, and uses are discussed. Transmutation and radioactive decay, as well as nuclear fission and fusion, are explained.

• **What activity is described in the textbook introduction?** (The making of soap and the blowing of bubbles.)
• **Was the activity easy to understand based on the presentation in the introduction?** (No, a simple activity sounded complicated. Although the terminology may be unfamiliar, the concepts are not.)
• **If the activity had been described as making soap and blowing bubbles, would it have been easier to understand?** (Yes.)
• **What might you conclude about chemistry based on the description?** (Chemistry is the study of basic processes. Chemical reactions are involved in many aspects of life.)

DISCOVERY ACTIVITY

Chemical Mysteries

Through the two-part Discovery Activity, students will learn that one factor that affects chemical reactions is heat, or temperature. The activity applies chemistry to everyday, ordinary situations and makes the study of chemistry relevant to students.

Chapter 1 ATOMS AND BONDING

SECTION	HANDS-ON ACTIVITIES
1–1 What Is Chemical Bonding? pages O12–O16 Multicultural Opportunity 1–1, p. O12 ESL Strategy 1–1, p. O12	**Student Edition** ACTIVITY (Doing): A Model of Energy Levels, p. O13 ACTIVITY BANK: Up in Smoke, p. O150 **Laboratory Manual** The Law of Definite Composition, p. O11 **Activity Book** CHAPTER DISCOVERY: Electrons and Energy Levels, p. O9 **Teacher Edition** A Different View of Salt, p. O10d Observing Bonding, p. O10d
1–2 Ionic Bonds pages O16–O20 Multicultural Opportunity 1–2, p. O16 ESL Strategy 1–2, p. O16	**Student Edition** ACTIVITY (Discovering): Steelwool Science, p. O18 ACTIVITY (Doing): Growing Crystals, p. O20
1–3 Covalent Bonds paes O20–O26 Multicultural Opportunity 1–3, p. O20 ESL Strategy 1–3, p. O20	**Student Edition** ACTIVITY (Doing): Bouncing Bonds, p. O21 LABORATORY INVESTIGATION: Properties of Ionic and Covalent Compounds, p. O32 **Laboratory Manual** Comparing Covalent and Ionic Compounds, p. O7 **Activity Book** ACTIVITY: Constructing Molecule Models, p. O23
1–4 Metallic Bonds pages O26–O28 Multicultural Opportunity 1–4, p. O26 ESL Strategy 1–4, p. O26	**Student Edition** ACTIVITY BANK: Hot Stuff, p. O151
1–5 Predicting Types of Bonds pages O28–O31 Multicultural Opportunity 1–5, p. O28 ESL Strategy 1–5, p. O28	**Student Edition** ACTIVITY BANK: The Milky Way, p. O152
Chapter Review pages O32–O35	

OUTSIDE TEACHER RESOURCES
Books
Boschke, F. L. (ed.) *Bonding and Structure,* Springer-Verlag.

Gray, Harry B. *Chemical Bonds: An Introduction to Atomic and Molecular Structure,* Benjamin-Cummings.

Hatfield, William E., and William E. Parker. *Symmetry in Chemical Bonding and Structure,* Merrill.

OTHER ACTIVITIES	MEDIA AND TECHNOLOGY
Activity Book ACTIVITY: Atomic Structure, p. O13 **Review and Reinforcement Guide** Section 1–1, p. O5	**Transparency Binder** Structure of the Atom Energy Levels **Courseware** Chemical Bonding (Supplemental) The Atomic Nucleus (Supplemental) Periodic Table (Supplemental) **Video/Videodisc** Chemical Bonding and Atomic Structure **English/Spanish Audiotapes** Section 1–1
Activity Book ACTIVITY: Bonding and Chemical Formulas, p. O29 ACTIVITY: Classifying Crystals, p. O15 **Review and Reinforcement Guide** Section 1–2, p. O7	**Transparency Binder** Ionic Bonding Periodic Table **Video/Videodisc** Periodic Table and Periodicity **English/Spanish Audiotapes** Section 1–2
Activity Book ACTIVITY: Electron-Dot Diagram, p. O17 ACTIVITY: Drawing Chemical Bonds, p. O19 **Review and Reinforcement Guide** Section 1–3, p. O9	**Transparency Binder** Covalent Bonding Molecules **English/Spanish Audiotapes** Section 1–3
Activity Book ACTIVITY: Bond Identification, p. O25 **Review and Reinforcement Guide** Section 1–4, p. O11	**English/Spanish Audiotapes** Section 1–4
Student Edition ACTIVITY (Calculating): Determining Oxidation Numbers, p. O29 ACTIVITY (Reading): Dangerous Foods, p. O30 **Activity Book** ACTIVITY: Charting Oxidation Number, p. O21 ACTIVITY: Chemical Analogies, p. O27 **Review and Reinforcement Guide** Section 1–5, p. O13	**English/Spanish Audiotapes** Section 1–5
Test Book Chapter Test, p. O9 Performance-Based Tests, p. O117	**Test Book** Computer Test Bank Test, p. O15

* All materials in the Chapter Planning Guide Grid are available as part of the Prentice Hall Learning System.

Murell, John N., et al. *The Chemical Bond,* Wiley.

Audiovisuals

Bonding Between Atoms of Different Elements: Metals and Non-Metals—The Ionic Bond, filmstrip or slides with cassette, Prentice-Hall Media

CHAPTER OVERVIEW

In order for us to understand the reasons behind the appearance and characteristics of the substances around us, we must understand how the basic units of matter—atoms—are attached to one another. Most atomic bonding, or "attaching," can be classified into three categories: ionic, covalent, and metallic. Each category describes a method by which atoms may bond to one another. By classifying a bonding arrangement, it is also immediately known that the substance has certain properties.

Ionic bonds typically are rigid, conduct electricity when melted or dissolved in water, and are likely to be composed of a metal and a nonmetal. Covalent bonds typically are composed of elements that are somewhat similar, sharing valence electrons, and frequently are gases or liquids. Metallic bonds are made up of one or more types of metal atoms, conducting electricity in solid form, and usually can be hammered into various shapes.

1–1 WHAT IS CHEMICAL BONDING?
THEMATIC FOCUS

In this section the concept that all substances in the universe are made up of essentially 109 different types of elements is presented. The combinations or arrangements of these 109 "universal" parts or ingredients produce the many substances that we see. A fact of major importance is that many of these combinations or arrangements are possible due to the mutual attraction of the outermost valence electrons in each atom to other atoms. These attractions for valence electrons usually produce a filled or complete energy level when two or more atoms bond.

The themes that can be focused on in this section are patterns of change, scale and structure, stability, and unity and diversity.

*Patterns of change: The combining of the atoms of elements to form new substances is called chemical bonding.

*Scale and structure: The atom consists of a nucleus (or center), in which are found protons and neutrons, and surrounding energy levels, in which are found electrons.

*Stability: In order to achieve stability, an atom will either gain or lose electrons and thereby will attain a complete outermost energy level.

Unity and diversity: Regardless of type, all bonds involve electrons—usually the valence electrons.

PERFORMANCE OBJECTIVES 1–1

1. Describe chemical bonding in terms of an atom's electron arrangement.
2. Define energy level.
3. List the maximum number of electrons in the first three energy levels.

SCIENCE TERMS 1–1
atom p. O12
chemical bonding p. O12
valence electron p. O13

1–2 IONIC BONDS
THEMATIC FOCUS

The purpose of this section is to introduce students to one of the primary ways by which atoms obtain a full, stable outermost energy level. Students will learn that some elements have only a small attraction for their own valence electrons, whereas other elements have a very strong attraction for their own valence electrons. In the latter case, the atom's attraction for electrons goes beyond its own electrons, extending to the point of taking electrons from other atoms in order to fill up its own outer energy level. This transfer of electrons will cause the atoms to have full, stable outer energy levels and to become charged or ionized. This type of bonding, involving charged atoms, is called ionic bonding.

The themes that can be focused on in this section are energy, patterns of change, and systems and interactions.

*Energy: Energy is always involved in chemical bonding. Ionization, or the process of removing electrons from an atom and forming an ion, requires energy (known as ionization energy). When an electron is gained, energy is released.

*Patterns of change: Ions of opposite charge strongly attract each other.

*Systems and interactions: Ionic bonding involves the transfer of electrons—one atom gains; the other atom loses—and the formation of ions and crystal lattices.

PERFORMANCE OBJECTIVES 1–2

1. Predict the resulting charge on an atom when electrons are added to or taken away from a specified atom.
2. Identify elements that have either low ionization energy or high electron affinity.
3. Describe the result of ionic bonding between elements as a regular pattern of ions in a crystal lattice.

SCIENCE TERMS 1–2
ionic bonding p. O16
ion p. O16
ionization p. O18
ionization energy p. O18
electron affinity p. O18
crystal lattice p. O19

1–3 COVALENT BONDS
THEMATIC FOCUS

One way for atoms to achieve the stability of a full energy level is to share their electrons with another atom. This section will explain to students that atoms with similar electron attracting abilities will very likely share their electrons with each other. The term *covalent* implies cooperation among the valence electrons. A major consequence of electron sharing is the formation of molecules.

The themes that can be focused on in this section are scale and structure and systems and interactions.

*Scale and structure: Network solids contain large numbers of atoms covalently bonded to one another.

*Systems and interactions: Covalent bonds involve the sharing of electrons and the formation of molecules.

PERFORMANCE OBJECTIVES 1–3

**1. Predict which atoms are most likely to engage in covalent bonding.
2. Compare and contrast the various characteristics of ionic crystals and covalent molecules.
3. Construct an electron-dot diagram for a covalently bonded molecule.**

SCIENCE TERMS 1–3

**covalent bonding p. O20
electron-dot diagram p. O21
diatomic element p. O22
molecule p. O23
network solid p. O24
polyatomic ion p. O24**

1–4 METALLIC BONDS
THEMATIC FOCUS

Metallic bonding is another way in which some atoms are able to bond to each other. In this type of bonding, which is explained to students in this section, metals can be seen to have a weak hold on their valence electrons. Although the attraction is generally weaker than that of the nonmetals, it is strong enough to be attractive to electrons. This situation allows the valence electrons of nearby atoms to "wander" to neighboring atoms, creating a mobile "sea of electrons."

The theme that can be focused on in this section is systems and interactions.

*Systems and interactions: Metallic bonding involves a sea of mobile electrons.

PERFORMANCE OBJECTIVES 1–4

**1. Define metallic bond.
2. Describe the properties of metals.
3. Identify which elements are likely to participate in metallic bonding.**

SCIENCE TERMS 1–4

metallic bond p. O26

1–5 PREDICTING TYPES OF BONDS
THEMATIC FOCUS

This section will explain to students that there are ways to predict various bonds using the periodic table of the elements. Students will discover that elements on the left of the periodic table give up their valence electrons while bonding, becoming positively charged as a result. This positive charge is called the oxidation number. Atoms on the right of the periodic table accept electrons, becoming negatively charged as a result. This negative charge is also known as the oxidation number. In a correctly written formula, the oxidation numbers expressed for all atoms should equal a net total of zero.

The theme that can be focused on in this section is stability.

*Stability: The sum of the oxidation numbers of the atoms in a compound must be zero.

PERFORMANCE OBJECTIVES 1–5

**1. Predict the common oxidation numbers of atoms, based on the position of that atom on the periodic table.
2. Predict the formation of compounds between elements.**

SCIENCE TERMS 1–5

oxidation number p. O29

Discovery Learning

TEACHER DEMONSTRATIONS
MODELING
A Different View of Salt

Without students' knowledge, put a spoonful of table salt into a beaker. Place an overhead projector very close to a viewing screen. Show students the white material in the beaker and turn the projector on, sprinkling some of the white material onto the stage of the projector. Then focus.
• **What do you see on the screen?** (Small objects.)
• **What shape do these objects have?** (Most students will say that the shapes are round.)

Move the projector 1 meter away from the screen and refocus.
• **What shape do the objects have?** (Most students will say that the shapes are round, although some might notice corners.)

Move the projector farther away from the screen and refocus to display a clear image.
• **What shape do the objects have?** (Most students will say that the objects appear to have square, or right-angle, corners.)

Place a large cube of halite (rock salt) in view of the students. Point out that both the small grains and the large cube are crystals of the same material. Explain that the material is salt, or the compound NaCl, and like most ionic compounds, salt forms crystals with right-angle corners.

Observing Bonding

To illustrate that substances bond in more than one way, use small samples of pure metals in a display, along with small samples of those same metals bonded with other elements in a compound. Some examples might be lead (metallic) and lead nitrate (ionic) or tin (metallic) and tin nitrate (ionic). Demonstrate that the metals will not dissolve in water but that the ionic compounds will. Point out the names of the respective bonds to students.

Also display a lead pencil and a diamond. Point out that both are pure carbon but, due to different bonding, have extremely different properties. Mention the names and the structure of the respective bonds to students.

CHAPTER 1
Atoms and Bonding

INTEGRATING SCIENCE

This physical science chapter provides you with numerous opportunities to integrate other areas of science, as well as other disciplines, into your curriculum. Blue numbered annotations on the student page and integration notes on the teacher wraparound pages alert you to areas of possible integration.

In this chapter you can integrate physical science and atoms (p. 12), life science and cells (p. 13), physical science and charge (pp. 13, 18), physical science and periodic table (p. 14), earth science and minerals (p. 20), language arts (pp. 20, 30), earth science and diamonds (p. 25), physical science and metals (p. 26), physical science and heat (p. 28), and mathematics (p. 29).

SCIENCE, TECHNOLOGY, AND SOCIETY/COOPERATIVE LEARNING

Chlorofluorocarbons are among the world's most widely used chemicals. These molecules, which contain chlorine, fluorine, and carbon atoms, are used in air conditioners and refrigerators throughout industry because they are stable, inflammable, and noncorrosive. Their stability, however, is creating a problem for the environment.

Chlorofluorocarbons, often called CFCs, do not break down in the lower layers of our atmosphere when released. These molecules float to the upper atmosphere, where intense ultraviolet light from the sun breaks the molecules apart, freeing chlorine atoms which attack the Earth's protective ozone layer. Free chlorine atoms bond with one of ozone's three oxygen atoms to create a chlorine-oxygen combination and an oxygen molecule. When the chlorine-oxygen molecule encounters a free oxygen atom, the chlorine atom gives up its oxygen atom to form a molecule of oxygen. The chlorine atom is then free to break down the next ozone molecule it encounters.

The Montreal Protocol, a treaty signed in 1987, was an agreement among 24 nations to cut production of the worst CFCs in half by 1998. More recent negotiations

INTRODUCING CHAPTER 1

DISCOVERY LEARNING

▶ *Activity Book*

Begin your teaching of the chapter by using the Chapter 1 Discovery Activity from the *Activity Book*. Using this activity, students will explore electrons and energy levels.

USING THE TEXTBOOK

Have students observe the picture on page O10.
• **What do you observe in the picture?** (Accept all answers.)
Have students read the caption to the picture on page O11.
• **What is shown in this picture?** (Salt crystals.)
• **Can you see any patterns in the shapes of these crystals?** (Accept all responses. Students will likely note that the crystals

Atoms and Bonding

Salt! You are surely familiar with this abundant and important chemical substance. As a frequently used seasoning, salt enhances the flavor of many of the foods you eat. Just imagine popcorn, chow mein, or collard greens without a dash of salt!

For a long time, salt has figured prominently in human affairs. Because it was both extremely important and limited in supply, salt was highly valued. Roman soldiers were, in fact, paid in cakes of salt. It is from the Latin word *sal*, meaning salt, that our word salary is derived.

Salt has come to symbolize many characteristics of human behavior. "The salt of youth" was William Shakespeare's description of the liveliness of this time of life. A person possessing valuable qualities is often described as being "worth one's salt." And those great human beings who have improved our world are often referred to as "the salt of the earth."

Salt is the substance sodium chloride. It is made of the elements sodium and chlorine. How and why do these elements combine to form salt? And what is the process by which hundreds of thousands of other substances are formed? Throw some salt over your left shoulder for luck and read on for the answers!

Journal *Activity*

You and Your World What do you think of when you hear the word element? How about the word compound? Do you know how these substances figure in your daily life? In your journal, answer these questions and provide any other thoughts you have about this topic. When you have completed this chapter, see if you need to modify your answers.

In ancient times, no other single substance equaled the importance of salt. Crystals of salt are shown in this magnified image.

O ■ 11

---

have been aimed at further reducing CFC production in an effort to protect the ozone layer and reduce global warming. Before a total ban can be agreed upon and enforced, scientists will have to find a safe alternative to this group of chemical compounds.

Cooperative learning: Using preassigned groups or randomly selected teams, have groups complete one of the following assignments:

- Automobile air conditioners are a major source of CFC pollution. Have groups respond to the following scenario: Your state legislature is considering an immediate ban on automobile air conditioners that use CFCs. Groups should present one of the following viewpoints at a public hearing on this proposed legislation:
  - Chemical industry (a good CFC substitute is three to five years away)
  - "Big 3" auto manufacturers (a major economic impact due to redesign of entire air-conditioning system)
  - Members of groups that represent automobile mechanics (the livelihood of some repair stations depends on automobile-air-conditioner recharge and repair)
  - An "average" person
  - An environmentalist

Encourage groups to consider and propose alternative legislation consistent with their assigned viewpoint.

- Have groups present the series of chemical reactions that explain how CFCs destroy the ozone layer. Group presentations could be illustrations with labels, posters, or the performance of a skit or soap opera using exaggerated emotions and reactions.

See Cooperative Learning in the *Teacher's Desk Reference.*

JOURNAL ACTIVITY

You may want to use the Journal Activity as the basis for a class discussion. If students have difficulty determining how these substances are a part of their daily lives, provide them with an example of an element and a compound or provide them with the definition of the terms. Then lead them in a discussion of how various elements and compounds might influence their daily lives. Students should be instructed to keep their Journal Activity in their portfolio.

---

appear to have flat surfaces and distinct corners.)
- **Can you see any patterns in the arrangement of these crystals?** (Accept all responses. Students might suggest that the patterns appear random.)
- **How do you think the atoms that form these crystals are held together?** (Accept all logical responses.)

Have students read the chapter introduction on page O11.
- **What word that we use is derived from *sal*, the Latin word meaning salt?** (Salary.)

- **Why were people, such as Roman soldiers, happy to be paid wages in the form of salt?** (Salt was important, in limited supply, and considered to be extremely valuable.)
- **Which two elements combine to form salt?** (Sodium and chlorine.)
- **How do you suppose sodium and chlorine are bonded, or held together, to form salt?** (Accept all answers. Point out that the answer to this question and others will be explored in this chapter.)

1-1 What Is Chemical Bonding?

MULTICULTURAL OPPORTUNITY 1-1

Assign students the task of building a wall-sized periodic table of the elements. Each student should be assigned a different element of the first 20 to 30 elements. Using large sheets of paper, they should research each element and draw a large Bohr atomic model of the element. Finally, display each of the elements, in its proper sequence, on the wall or large bulletin board.

ESL STRATEGY 1-1

Ask students to draw a diagram of an atom. Have them label the nucleus, its charge, and its subatomic particles. Students should also label the electron cloud, its charge, its levels, and should indicate the number of electrons each level carries.

TEACHING STRATEGY 1-1

FOCUS/MOTIVATION

It should be pointed out to students at this time that electrons in atoms have some restrictions. To do this, you might remind them of some restrictions that have been a part of their life experiences by discussing questions similar to the following: Have you ever flown on a large jet airplane? Did you notice that some seating sections are limited to two seats per row, whereas in another section four or even more seats might be in one row?

A similar analogy for discussion is a resort hotel that has some rooms that have occupancy for two persons and other rooms that hold four or six persons. Have students describe other examples from their lives in which they have experienced restrictions analogous to the restrictions imposed on electrons.

Point out that the electron energy levels in atoms are also restricted and that each level can hold only certain numbers of electrons.

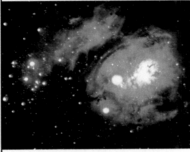

Figure 1–1 *The hundreds of thousands of different substances in nature are made up of atoms. Atoms are the basic building blocks of everything in the universe, including all forms of living things, the air, and the Lagoon Nebula in the constellation Sagittarius.*

12 ■ O

1-1 What Is Chemical Bonding?

Look around you for a moment and describe what you see. Do you see the pages of this textbook? A window? Perhaps trees or buildings? Your friends and classmates? The air you breathe? All these objects—and many others not even mentioned—have one important property in common. They are all forms of matter. And all matter—regardless of its size, shape, color, or phase—is made of tiny particles called **atoms**. Atoms are the basic building blocks of all the substances in the universe. As you can imagine, there are hundreds of thousands of different substances in nature. (To prove this, try to list all the different substances you can. Is there an end to your list?) ❶

Yet, as scientists know, there are only 109 different elements. Elements are the simplest type of substance. Elements are made of only one kind of atom. How can just 109 different elements form so many different substances? ❷

The 109 elements are each made of specific type of atoms. Atoms of elements combine with one another to produce new and different substances called compounds. You are already familiar with several compounds: water, sodium chloride (table salt), sugar, carbon dioxide, vinegar, lye, and ammonia. Compounds contain more than one kind of atom chemically joined together.

The combining of atoms of elements to form new substances is called chemical bonding. Chemical bonds are formed in very definite ways. The atoms combine according to certain rules. The rules of **chemical bonding** are determined by the structure of the atom, which you are now about to investigate.

Electrons and Energy Levels

The atom contains a positively charged center called the nucleus. Found inside the nucleus are two types of subatomic (smaller than the atom) particles protons and neutrons. Protons have a positive charge, and neutrons have no charge. Neutrons are

---

CONTENT DEVELOPMENT

An important idea concerning electrons is the stability of an octet, or a full energy level. Ask students to suppose that they were in a rowboat that had a rowing capacity of eight people. The arrangement of people rowing the boat might be four rows of two each. But if the boat had only seven people rowing, point out that it would need one more person for maximum efficiency and stability. Use this analogy, and others developed by students, to help discuss and understand the importance of a balanced number of electrons in energy levels.

● ● ● ● **Integration** ● ● ● ●

Use the idea of the basic building blocks of all substances to integrate concepts of the atom into your lesson.

Use Figure 1–2 to integrate concepts of cells into your lesson.

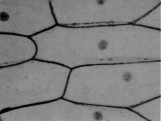
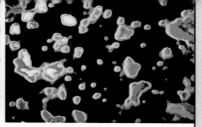

Figure 1–2 *You can think of atoms as similar to the individual* ②
*stones that make up the pyramids in Giza, Egypt, or the individual
cells of which all living things are made. The photograph of uranium
atoms taken with an electron microscope gives you some idea of
what atoms actually look like—magnified more than five million
times, that is!*

neutral particles. Thus the nucleus as a whole has a positive charge.

Located outside the nucleus are negatively charged particles called electrons. The negative charge of the electrons balances the positive charge of the nucleus. The atom as a whole is neutral. It has no net charge. What do you think this means about the number of electrons (negatively charged) compared with the number of protons (positively charged)? ③

The negatively charged electrons of an atom are attracted by the positively charged nucleus of that atom. This electron-nucleus attraction holds the atom together. The electrons, however, are not pulled into the nucleus. They remain in a region outside the nucleus called the electron cloud.

The electron cloud is made up of a number of different energy levels. Electrons within an atom are arranged in energy levels. Each energy level can hold only a certain number of electrons. The first, or innermost, energy level can hold only 2 electrons, the second can hold 8 electrons, and the third can hold 18 electrons. The electrons in the outermost energy level of an atom are called **valence electrons.** The valence electrons play the most significant role in determining how atoms combine.

When the outermost energy level of an atom contains the maximum number of electrons, the level is full, or complete. Atoms that have complete (filled) outermost energy levels are very stable. They usually do not combine with other atoms to form compounds. They do not form chemical bonds.

ACTIVITY
DOING

A Model of Energy Levels

1. Cut a thin piece of corkboard into a circle 50 cm in diameter to represent an atom.

2. Insert a colored pushpin or tack into the center to represent the nucleus.

3. Draw three concentric circles around the nucleus to represent energy levels. The inner circle should be 20 cm in diameter; the second circle, 30 cm in diameter; and the third, 40 cm in diameter.

4. Using pushpins or tacks of another color to represent electrons, construct the following atoms: hydrogen (H), helium (He), lithium (Li), fluorine (F), neon (Ne), sodium (Na), and argon (Ar).

Are any of these elements in the same family? If so, which ones? How do you know?

O ■ 13

ACTIVITY
DOING
A MODEL OF ENERGY LEVELS

Skills: Making models, applying concepts
Materials: corkboard, scissors, pushpins or tacks of different colors

In this activity students reinforce their understanding of energy levels by constructing models of various atoms and their electron configurations. Check students' models carefully to ensure that they have placed the correct number of electrons in each atom's energy levels. Hydrogen, lithium, and sodium are in the same family as are helium, neon, and argon. You can tell based on the number of valence electrons that they have.

INDEPENDENT PRACTICE
📖 Media and Technology

Use the transparency called Structure of the Atom to help develop the concept of atomic structure.

CONTENT DEVELOPMENT

Remind students about some of the concepts of atoms by pointing out that whether someone is building a home or a very large building, cement will be used to bond the bricks and blocks of the building together.

Remind students that atoms are our building blocks. Instead of being held together by cement, the atoms of our bodies are held together by chemical bonds. These bonds are dependent on electrons. Electrons interlock, or bond together, various atoms. The kinds of atoms that are able to be bonded together depend on the numbers and locations of the electrons within atoms. Stress to students that the valence electrons, or the electrons in the outermost energy level of an atom, play the most significant role in determining how atoms combine, or bond.

● ● ● ● **Integration** ● ● ● ●

Use the discussion of the electrical nature of protons, neutrons, and electrons to integrate charge concepts into your lesson.

The subatomic world is an invisible world to the naked eye. Have interested students use research materials to discover the actual sizes of subatomic particles such as protons, neutrons, and electrons. Challenge students to use their findings to develop comparative size analogies of the particles (for example, if one of the subatomic particles were the size of a basketball, another would be the size of a pea). Have successful students share their findings with the class.

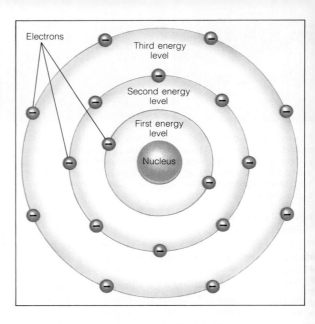

Figure 1–3 *An atom contains a positively charged nucleus surrounded by negatively charged electrons located in energy levels within the electron cloud. Each energy level can hold only a certain number of electrons. How many electrons can the first energy level hold? The second? The third?* ❶

Activity Bank

Up in Smoke, p.150

Turn to Appendix E on pages 166–167 and study it carefully. You are looking at the periodic table of the elements—one of the most important "tools" of a physical scientist. All the known elements (109) are listed in this table in a specific way. Every element belongs to a family, which is a numbered, vertical column. There are 18 families of elements. Every element also belongs to a period, which is a ❶ numbered, horizontal row. How many periods of ele-❷ ments do you see? Elements in the same family have similar properties, the most important of which is the number of electrons in the outermost energy level, or the number of valence electrons.

Look at Family 18 in the periodic table. It contains the elements helium, neon, argon, krypton, xenon, and radon. The atoms of these elements do not form chemical bonds under normal conditions. This is because all the atoms of elements in Family 18 have filled outermost energy levels. Remember, if the first energy level is also the outermost, it needs only 2 electrons to make it complete. Can you tell which element in Family 18 has only 2 valence electrons? ❸

1–1 (continued)

INDEPENDENT PRACTICE

📖 **Media and Technology**

Use the transparency in the *Transparency Binder* called Energy Levels to help develop the concept of electrons and energy.

GUIDED PRACTICE

Skills Development

Skill: Interpreting diagrams

Have students observe Figure 1–3 and remind them that energy levels in atoms are restricted and can hold only certain numbers of electrons. Ask:

• **How many electrons can the first, or innermost, energy level hold?** (2.)
• **How many electrons can the second energy level hold?** (8.)

• **How many electrons can the third energy level hold?** (8.)
• **Which energy level represents the valence electrons?** (The third energy level.)
• **Would you predict that the atom pictured in this diagram is stable or unstable?** (Lead students to suggest that it is stable.)
• **Why?** (Lead students to recall that if the outermost energy level of an atom is filled to capacity, the atom is quite stable. The capacity of the third energy level of any atom is 8 electrons, and the outermost, or third, energy level of the atom

in this diagram has 8 electrons and is filled to capacity.)

CONTENT DEVELOPMENT

Examine the periodic table and point out that Family 18 contains elements that do not form chemical bonds under normal conditions.

• **Which elements are found in Family 18 of the periodic table?** (Helium, neon, argon, krypton, xenon, and radon.)

Point out that because of their nonreactive nature, these gases have many uses.

18
2 **He** Helium 4.003
10 **Ne** Neon 20.179
18 **Ar** Argon 39.948
36 **Kr** Krypton 83.80
54 **Xe** Xenon 131.29
86 **Rn** Radon (222)

Figure 1–4 *These balloons are filled with helium, a highly unreactive element. Neon gas, another unreactive element, is used in neon lights. Argon, shown here as a laser made visible through smoke, is also highly unreactive. These three elements are members of Family 18. What must be true of the outermost energy levels of atoms in this family?* ④

Electrons and Bonding

The electron arrangement of the outermost energy level of an atom determines whether or not the atom will form chemical bonds. As you have just read, atoms of elements in Family 18 have complete outermost energy levels. These atoms generally do not form chemical bonds.

Atoms of elements other than those in Family 18 do not have filled outermost energy levels. Their outermost energy level lacks one or more electrons to be complete. Some of these atoms tend to gain electrons in order to fill the outermost energy level. Fluorine (F), which has 7 valence electrons, gains 1 electron to fill its outermost energy level. Other atoms tend to lose their valence electrons and are left with only filled energy levels. Sodium (Na), which has 1 valence electron, loses 1 electron.

Figure 1–5 *The outermost energy level of a helium atom contains the maximum number of electrons—2. In a neon atom, the outermost energy level is the second energy level. It contains the maximum 8 electrons. What chemical property do these elements share?* ⑤

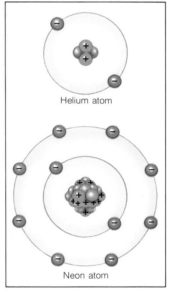

Helium atom

Neon atom

O ■ 15

Some uses include arc welding, food processing (the metal and foods cannot oxidize), and in light bulbs (to allow high temperatures to be generated without breaking down the tungsten filaments.)

● ● ● ● **Integration** ● ● ● ●

Use the introduction of element families to integrate concepts of the periodic table into your lesson.

1-2 Ionic Bonds

MULTICULTURAL OPPORTUNITY 1-2

Have students build models of elements that can combine by ionic bonds. Sodium chloride might be a good example. Students can make Bohr atoms using sheets of paper and gummed dots to represent the electrons. Combining the diagrams for sodium with that of chlorine should make it clear to students how an ionic bond is formed.

ESL STRATEGY 1-2

Write the following activities on the chalkboard and have ESL students work with English-speaking partners in preparing their answers.
1. State the kind of action that occurs in ionic bonding.
2. Give the definition and synonym for *electron-transfer bonding*.
3. Describe the ionization process and the type of energy it requires.

Illustrate an ordinary lattice; then ask students if they think the term *crystal lattice* is appropriate. Have students support their answer. If they think crystal lattice is inappropriate, ask them to suggest a different name.

1-1 (continued)

INDEPENDENT PRACTICE
Media and Technology

Show the Video/Videodisc called Chemical Bonding and Atomic Structure to introduce the concepts of metallic, covalent, and ionic bonds. The video explores electrostatic forces; the behavior of valence electrons in metallic, covalent, and ionic bonding; energy changes resulting from the making and breaking of bonds; and the structure of atomic, ionic, and molecular crystals. After viewing the video, students should choose a common alloy, research its composition, and predict the configuration of its bond.

In order to achieve stability, an atom will either gain or lose electrons. In other words, an atom will bond with another atom if the bonding gives both atoms complete outermost energy levels. In the next section you will learn how bonding takes place.

1-1 Section Review

1. What is chemical bonding?
2. What is the basic structure of the atom?
3. What are valence electrons? How many valence electrons can there be in the first energy level? In the second? In the third?
4. What determines whether or not an atom will form chemical bonds?

Connection—History
5. On May 6, 1937, the airship (blimp) *Hindenburg* exploded in midair just seconds before completing its transatlantic voyage. On board the blimp had been 210,000 cubic meters of hydrogen gas. Since that time, airships have used helium gas rather than hydrogen gas to keep them aloft. Explain why.

Guide for Reading
Focus on this question as you read.
▶ What is ionic bonding?

Figure 1–6 *Bonding usually results in the formation of compounds, such as ammonium chloride.*

1-2 Ionic Bonds

As you have just learned, an atom will bond with another atom in order to achieve stability, which means in order for both atoms to get complete outermost energy levels. One way a complete outermost energy level can be achieved is by the transfer of electrons from one atom to another. Bonding that involves a transfer of electrons is called **ionic bonding**. Ionic bonding, or electron-transfer bonding, gets its name from the word **ion**. An ion is a charged atom. Remember, an atom is neutral. But if there is a transfer of electrons, a neutral atom will become a charged atom.

Because ionic bonding involves the transfer of electrons, one atom gains electrons and the other atom loses electrons. Within each atom the negative and positive charges no longer balance. The atom that has gained electrons has gained a negative

ENRICHMENT

Actually, some noble gases have been involved in bonding. For example, xenon has formed compounds from reaction with fluorine. Challenge interested students with the task of discovering how this was accomplished.

INDEPENDENT PRACTICE
Section Review 1-1

1. The combining of atoms of elements to form new substances.

2. Atoms contain a positively charged center called the nucleus, which contains electrically positive protons and electrically neutral neutrons. Negatively charged electrons are located outside the nucleus in the electron cloud.

3. Electrons in the outermost energy level of an atom: 2; 8; 8.

4. The number of valence electrons that it has.

5. Helium is less dense than air and is nonreactive with most substances, especially oxygen, and heat and sparks.

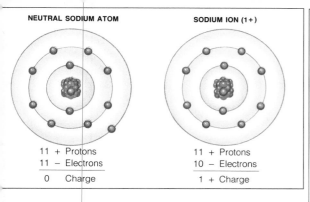

NEUTRAL SODIUM ATOM	SODIUM ION (1+)
11 + Protons	11 + Protons
11 − Electrons	10 − Electrons
0 Charge	1 + Charge

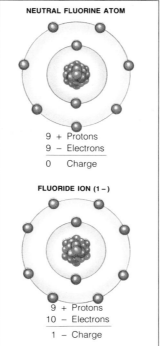

NEUTRAL FLUORINE ATOM

9 + Protons
9 − Electrons
0 Charge

FLUORIDE ION (1−)

9 + Protons
10 − Electrons
1 − Charge

charge. It is a negative ion. For example, fluorine (F) has 7 valence electrons. To complete its outermost energy level, the fluorine atom gains 1 electron. In gaining 1 negatively charged electron, the fluorine atom becomes a negative ion. The symbol for the fluoride ion is F^{1-}. (For certain elements, the name of the ion is slightly different from the name of the atom. The difference is usually in the ending of the name—as with the fluorine atom and the fluoride ion.)

The sodium atom (Na) has 1 valence electron. When a sodium atom loses this valence electron, it is left with an outermost energy level containing 8 electrons. In losing 1 negatively charged electron, the sodium atom becomes a positive ion. The symbol for the sodium ion is Na^{1+}. Figure 1–7 shows the formation of the fluoride ion and the sodium ion.

Figure 1–8 *The general rule that opposites attract is responsible for the formation of the ionic bond between a positive sodium ion and a negative fluoride ion. Notice the transfer of an electron during the ionic bonding. What is the formula for the resulting compound?* ②

Figure 1–7 *The formation of a negative fluoride ion involves the gain of an electron by a fluorine atom. The formation of a positive sodium ion involves the loss of an electron by a sodium atom. How many valence electrons does a fluorine atom have? A sodium atom? What is the symbol for a fluoride ion? A sodium ion?* ①

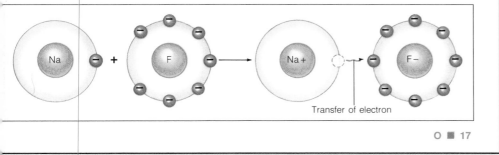

Transfer of electron

O ■ 17

REINFORCEMENT/RETEACHING

Monitor students' responses to the Section Review questions. If students appear to have difficulty with any of the questions, review the appropriate material in the section.

CLOSURE

▶ *Review and Reinforcement Guide*

At this point have students complete Section 1–1 in the *Review and Reinforcement Guide.*

TEACHING STRATEGY 1–2

FOCUS/MOTIVATION

An ionic substance that most students have had experience with is table salt—NaCl. Obtain a large container of table salt. In front of the class, slowly pour the salt from the container into a beaker that is located about one-third meter below the container of salt. While the salt is being poured, ask:

• **What are some things you know about salt?** (Students might suggest that it dissolves in water, it is made of sodium and chlorine, it tastes "salty," too much of it may be unhealthy, and so forth.)

Ask students to locate sodium and chlorine on the periodic table. Explain that the known formula for salt is NaCl.

CONTENT DEVELOPMENT

Point out that atoms can bond in many different combinations and arrangements. Remind students that although an atom is electrically neutral, when it gains or loses electrons, it becomes a charged atom, or an ion.

INDEPENDENT PRACTICE

▥ **Media and Technology**

Use the transparency called Ionic Bonding to help develop the concept of electron transfer.

ENRICHMENT

▶ *Activity Book*

Students will be challenged by the Chapter 1 activity in the *Activity Book* called Bonding and Chemical Formulas. In it students will determine the charge of an ion after an exchange of electrons.

18 ■ O

ACTIVITY
DISCOVERING
STEELWOOL SCIENCE

Discovery Learning

Skills: Making observations, making comparisons

Materials: steelwool, water, 2 jars and 2 dishes of equal size and shape

In this investigation students should observe that the water level in each jar decreased relative to the amount of evaporation and condensation that was occurring in each jar. The greatest decrease, however, was shown in the jar containing steelwool. The moisture in the jar containing steelwool caused the steelwool to begin oxidizing, or rusting. Students should infer that the water and the steelwool reacted and that the purpose of the empty jar was to serve as a control.

1–2 (continued)

CONTENT DEVELOPMENT

Use the chalkboard to write examples that show all Group 1 metals form one-to-one ratios with Cl, LiCl, NaCl, KCl, and so on. Have volunteer students use the chalkboard to show the valence levels of Li, Na, and K. Then ask:

• **How many valence electrons are found in Group 1 metals?** (1.)

• **Would it be easier for those metals to attract 7 new valence electrons from other elements, or would it be easier to have some other element take the 1 outer level electron?** (It would be easier to have another element take the 1 outer level electron.)

• **Why?** (When the Group 1 metal loses its outer electron, the former next-to-outermost level is now the outermost level, and it is full.)

Point out that elements which have nearly a full outer level are likely to gain 1 or 2 electrons (rather than give some up) to complete that outer level. (Columns 6 and 7 on the periodic table qualify for this distinction.) Stress to stu-

---

ACTIVITY
DISCOVERING

Steelwool Science

1. Place a small amount of steelwool in a jar and push it to the bottom. Pack it tightly enough so that it will remain in place when the jar is inverted.

2. Into each of two identical dishes, pour water to a height of about 2 cm. Make sure there are equal amounts of water in each dish.

3. In one dish, place the jar with the steelwool mouth down in the water. In the other dish, place an identical but empty jar in the same position.

4. Observe the jars every day for 1 week. Record your observations.

What changes did you observe in the steelwool? In the water levels?

■ Explain your observations.

■ What is the purpose of the empty jar?

Figure 1–9 *During ionization, an electron is removed from an atom and an ion forms. Energy is absorbed during ionization. Energy is released when an atom gains an electron and forms an ion. What is the tendency of an atom to gain electrons called?* ③

In nature, it is a general rule that opposites attract. Since the two ions Na^{1+} and F^{1-} have opposite charges, they attract each other. The strong attraction holds the ions together in an ionic bond. The formation of the ionic bond results in the formation of the compound sodium fluoride, NaF. See Figure 1–8 on page 17.

Energy for Ion Formation

In order for the outermost electron to be removed from an atom, the attraction between the negatively charged electron and the positively charged nucleus must be overcome. The process of removing electrons and forming ions is called **ionization**. Energy is needed for ionization. This energy is called **ionization energy**.

The ionization energy for atoms that have few valence electrons is low. Do you know why? Only a small amount of energy is needed to remove electrons from the outermost energy level. As a result, these atoms tend to lose electrons easily and to become positive ions. What elements would you expect would have low ionization energies? ①

The ionization energy for atoms with many valence electrons is very high. These atoms do not lose electrons easily. As a matter of fact, these atoms usually gain electrons. It is much easier to gain 1 or 2 electrons than to lose 7 or 6 electrons! The tendency of an atom to attract electrons is called **electron affinity**. Atoms such as fluorine are said to have a high electron affinity because they attract electrons easily. What other atoms have a high electron affinity? ②

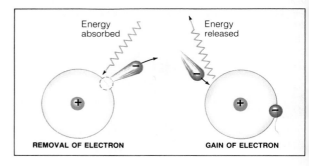

REMOVAL OF ELECTRON   Energy absorbed   Energy released   **GAIN OF ELECTRON**

---

dents that the consequence of giving up or obtaining additional electrons is to become charged. Electrons have a negative charge—when an atom gains an electron, the atom becomes negative.

INDEPENDENT PRACTICE
📖 **Media and Technology**

Use the transparency in the *Transparency Binder* called Periodic Table to reinforce understanding of the periodic relationships among the elements.

CONTENT DEVELOPMENT

Use the chalkboard to write several examples listing the proton and electron count of various neutral atoms. Then change the electron count by adding to the nonmetals and subtracting from the metals. Ask students to then predict ionic charges.

Arrangement of Ions in Ionic Compounds

Ions of opposite charge strongly attract each other. Ions of like charge strongly repel each other. As a result, ions in an ionic compound are arranged in a specific way. Positive ions tend to be near negative ions and farther from other positive ions.

The placement of ions in an ionic compound results in a regular, repeating arrangement called a **crystal lattice**. A crystal lattice is made of huge numbers of ions. A crystal lattice gives the compound great stability. It also accounts for certain physical properties. For example, ionic solids tend to have high melting points. Figure 1–10 shows the crystal lattice structure of sodium chloride.

Ionic compounds are made of nearly endless arrays of ions. A chemical formula for an ionic compound shows the ratio of ions present in the crystal lattice. It does not show the actual number of ions.

Each ionic compound has a characteristic crystal lattice arrangement. This lattice arrangement gives a particular shape to the crystals of the compound. For example, sodium chloride forms cubic crystals. Figure 1–11 shows some unusual shapes that result from ionic bonding.

Figure 1–11 *Ionic crystals often have unusual and amazingly beautiful shapes. Here you see ice crystals (left), crystals of the mineral rhodonite (center), and crystals of the mineral quartz (right).*

Cl⁻   Na⁺

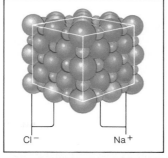

Figure 1–10 *Ionic bonding results in the formation of crystals. Crystals have a characteristic crystal lattice, or regular, repeating arrangement of ions. The lattice arrangement of sodium chloride crystals gives them their characteristic cubic shape. What is the common name for this crystal?* 4

 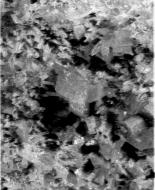 

O ■ 19

HISTORICAL NOTE
MEASURING CRYSTALS

William Bragg (1862–1942) and his son Lawrence worked with X-rays and the diffraction of these rays. They showed that X-rays would diffract when passed through crystals and that the measurement of this diffraction could be used to calculate the distance between the ions in the crystal. This technique for measuring ionic radii was similar to that used in the discovery of the bonding pattern and double-helix arrangement of the DNA molecule.

pleted, allow the entire class the opportunity to hear the questions and debate the correct answers.

GUIDED PRACTICE

Skills Development

Skill: Making diagrams

Have students draw the electrons in the energy levels of magnesium, noting that there are 2 electrons in the valence level. Then have them draw the electrons in the energy levels for oxygen, noting that there are 6 electrons in the valence level. They should then deduce the formula of magnesium oxide (MgO) and repeat the activity for calcium oxide (CaO) and calcium sulfide (CaS).

● ● ● ● **Integration** ● ● ● ●

Use the discussion of ion formation to integrate charge concepts into your lesson.

INDEPENDENT PRACTICE

🔍 **Media and Technology**

You may want to show the video/videodisc called Periodic Table and Periodicity to reinforce the concepts of elements and to extend the understanding of their relationships. The video explores the patterns and relationships of elements and their properties and explains why there are families of elements and gradual changes in the properties of elements arranged by atomic number across the periodic table. After viewing the video, have students work in groups to design a simple ten-question quiz about the periodic table. Groups should exchange quizzes and attempt to complete another group's quiz concerning the periodic table. After the quizzes have been com-

1-3 Covalent Bonds

MULTICULTURAL OPPORTUNITY 1-3

Have students build models of elements that can combine by covalent bonds. Have them build large electron-dot models for various molecules.

ESL STRATEGY 1-3

Have students describe the type of action that occurs in covalent bonding and compare it to the action that occurs in ionic bonding.

1-2 (continued)

● ● ● ● **Integration** ● ● ● ●

Use the example of classifying minerals by the shape of their crystals to integrate concepts of minerals into your lesson.

REINFORCEMENT/RETEACHING

▶ *Activity Book*

Students who need practice classifying ionic compounds should be provided with the Chapter 1 activity called Classifying Crystals.

INDEPENDENT PRACTICE

Section Review 1-2

1. Ionic bonding involves the transfer of electrons in which one atom gains electrons and one atom loses electrons.

2. The atom gains 1 or more electrons; the atom loses 1 or more electrons.

3. The energy required to remove an electron from a neutral atom; the tendency of an atom to attract electrons.

4. The placement of ions in an ionic compound; the chemical formula determines the ratio of ions present in the crystal lattice.

5. Accept logical examples. Students might suggest the opposite poles of a magnet.

REINFORCEMENT/RETEACHING

Review students' responses to the Section Review questions. Reteach any material that is still unclear, based on students' responses.

ACTIVITY: DOING

Growing Crystals

1. Make a small sliding loop in the end of a 10-cm length of thin plastic fishing line. Attach the loop to a small crystal of sea salt (NaCl).

2. In a beaker containing 200 mL of very hot water, stir to dissolve as much sea salt as the water will hold.

3. Suspend the fishing line containing the loop and crystal in the beaker. Tie the other end of the line to a pencil. Lay the pencil across the top of the beaker to support the line. The crystal should be suspended about halfway down in the liquid.

4. Observe the growing crystals every day for 3 days.

Guide for Reading

Focus on these questions as you read.

▶ What is covalent bonding?

▶ What is a molecule?

The crystal shape of an ionic compound is of great importance to geologists in identifying minerals. There are more than 2000 different kinds of minerals, and many of them look alike! One of the properties by which minerals are classified is crystal shape. There are six basic crystal shapes, or systems, and each of the thousands of minerals belongs to one of these systems.

1-2 Section Review

1. What is ionic bonding?
2. How does an atom become a negative ion? A positive ion?
3. What is ionization energy? Electron affinity?
4. What is a crystal lattice? What is the relationship between a chemical formula for an ionic compound and its crystal lattice?

Critical Thinking—*You and Your World*

5. A general rule in nature is that opposites attract. In addition to the behavior of oppositely charged ions, what other example(s) of this rule can you think of?

1-3 Covalent Bonds

Bonding often occurs between atoms that have high ionization energies and high electron affinities. In other words, neither atom loses electrons easily, but both atoms attract electrons. In such cases, there can be no transfer of electrons between atoms. What there can be is a sharing of electrons. Bonding in which electrons are shared rather than transferred is called **covalent bonding.** Look at the word covalent. Do you see a form of a word you have just learned? Do you know what the prefix *co-* means? Why is covalent an appropriate name for such a bond? ❶

By sharing electrons, each atom fills up its outermost energy level. So the shared electrons are in the outermost energy level of both atoms at the same time.

CLOSURE

▶ *Review and Reinforcement Guide*

Students may now complete Section 1-2 in the *Review and Reinforcement Guide.*

TEACHING STRATEGY 1-3

FOCUS/MOTIVATION

To point out that some substances do not have the same characteristics as the ionic materials in the previous section,

---

place some ionic salt crystals on one side of a double-pan balance. Place a balanced amount of alcohol (or similar fast-evaporating substance) on the other side of the balance. **CAUTION:** *Alcohol is extremely flammable—have no flame near this demonstration.* Within a short time, the materials will become unbalanced. Ask:

• **What did you observe?** (Prompt the class for the response: The alcohol is evaporating, causing less mass to be present. When this response is given, point out that covalent substances are made up of units

Figure 1–12 *Covalent bonding is bonding in which electrons are shared rather than transferred. Two substances that exhibit covalent bonding are sulfur (left) and sugar (right). Which substance is an element? A compound?* ②

Nature of the Covalent Bond

In covalent bonding, the positively charged nucleus of each atom simultaneously (at the same time) **attracts the negatively charged electrons that are being shared.** The electrons spend most of their time between the atoms. The attraction between the nucleus and the shared electrons holds the atoms together.

The simplest kind of covalent bond is formed between two hydrogen atoms. Each hydrogen atom has 1 valence electron. By sharing their valence electrons, both hydrogen atoms fill their outermost energy level. Remember, the outermost energy level of a hydrogen atom is complete with 2 electrons. The two atoms are now joined in a covalent bond. See Figure 1–13 on page 22.

Chemists represent the electron sharing that takes place in a covalent bond by an **electron-dot diagram.** In such a diagram, the chemical symbol for an element represents the nucleus and all the inner energy levels of the atom—that is, all the energy levels except the outermost energy level, which is the energy level with the valence electrons. Dots surrounding the symbol represent the valence electrons.

A hydrogen atom has only 1 valence electron. An electron-dot diagram of a hydrogen atom would look like this:

H·

---

that do not attract each other with much force, which allows them to separate.)

● ● ● ● **Integration** ● ● ● ●

Use the introduction of the term *covalent* to integrate language arts skills into your science lesson.

GILBERT LEWIS

Gilbert Lewis (1875–1946) published a paper entitled "The Atom and the Molecule" in 1916. In this paper, the idea of sharing the electrons among the atoms of nonionic compounds was first proposed. His research about electron pairs in molecular structure gave us the concept of the covalent bond. To illustrate the arrangement, Lewis provided the idea of what are now known as the "Lewis electron-dot structures," otherwise known as electron-dot formulas.

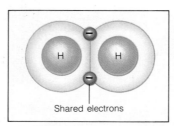

Figure 1–13 *The covalent bond between 2 atoms of hydrogen results in a molecule of hydrogen. In a covalent bond, the electrons are shared. How many valence electrons does each hydrogen atom have?* ❶

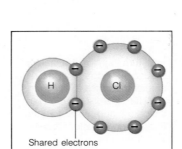

Shared electrons

The covalent bond between the two hydrogen atoms shown in Figure 1–13 can be represented in an electron-dot diagram like this:

H : H

The two hydrogen atoms are sharing a pair of electrons. Each hydrogen atom achieves a complete outermost energy level (an energy level containing 2 electrons).

Chlorine has 7 valence electrons. An electron-dot diagram of a chlorine atom looks like this:

: Cl ·

The chlorine atom needs one more electron to complete its outermost energy level. If it bonds with another chlorine atom, the two atoms could share a pair of electrons. Each atom would then have 8 electrons in its outermost energy level. The electron-dot diagram for this covalent bond would look like this:

: Cl : Cl :

Covalent bonding often takes place between atoms of the same element. In addition to hydrogen and chlorine, the elements oxygen, fluorine, bromine, iodine, and nitrogen bond in this way. These elements are called **diatomic elements**. When found in nature, diatomic elements always exist as two atoms covalently bonded.

The chlorine atom, with its 7 valence electrons, can also bond covalently with an unlike atom. For example, a hydrogen atom can combine with a chlorine atom to form the compound hydrogen chloride. See Figure 1–14. The electron-dot diagram for this covalent bond is

H : Cl :

You can see from this electron-dot diagram that by sharing electrons, each atom completes its outermost energy level.

Figure 1–14 *By sharing their valence electrons, hydrogen and chlorine form a molecule of the compound hydrogen chloride. When dissolved in water, hydrogen chloride forms the acid known as hydrochloric acid. Hydrochloric acid is an important part of the digestive juices found in the stomach.*

22 ■ O

1–3 (continued)

CONTENT DEVELOPMENT

Have students observe Figure 1–13. Ask:

• **In this diagram, what is the relationship between the valence electrons of these two hydrogen atoms?** (The valence electrons are being shared.)

Point out that when valence electrons from two different atoms are shared, the bond that results is a covalent bond. The shared electrons are in the outermost energy level of both atoms.

Explain that in a covalent bond, neither atom is weak enough to give up its electron—the strength of the bond comes from the attraction of the positively charged nuclei for the negatively charged electrons.

INDEPENDENT PRACTICE

📖 **Media and Technology**

Use the transparency in the *Transparency Binder* called Covalent Bonding to help develop the concept of bonding through sharing electrons.

GUIDED PRACTICE

Skills Development

Skills: Making observations, making comparisons

At this point have students complete the in-text Chapter 1 Laboratory Investigation: Properties of Ionic and Covalent Compounds. In this investigation students will discover whether or not differences exist between the properties of covalent and ionic compounds.

REINFORCEMENT/RETEACHING

To help reinforce the differences between types of bonds, ask the following question:

• **How is an ionic bond different from a covalent bond?** (Ionic bonding involves the transfer of electrons—one atom gains electrons, and another atom loses electrons. Covalent bonding involves bonding in which electrons are shared rather than transferred.)

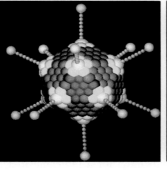
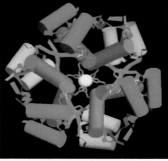

Figure 1–15 *Not all molecules are as simple as hydrogen chloride. Here you see computer-generated images of several more complex molecules: morphine (left), a common cold virus (center), and the hormone insulin (right).*

Formation of Molecules

In a covalent bond, a relatively small number of atoms are involved in the sharing of electrons. The combination of atoms that results forms a separate unit rather than the large crystal lattices characteristic of ionic compounds.

The combination of atoms formed by a covalent bond is called a **molecule** (MAHL-ih-kyool). A molecule is the smallest particle of a covalently bonded substance that has all the properties of that substance. This means that 1 molecule of water, for example, has all the characteristics of a glass of water, a bucket of water, or a pool of water. But if a molecule of water were broken down into atoms of its elements, the atoms would not have the same properties as the molecule.

Molecules are represented by chemical formulas. Like a chemical formula for an ionic crystal, the chemical formula for a covalent molecule contains the symbol of each element involved in the bond. Unlike a chemical formula for an ionic crystal, however, the chemical formula for a molecule shows the exact number of atoms of each element involved in the bond. The subscripts, or small numbers placed to the lower right of the symbols, show the number of atoms of each element. When there is only 1 atom of an element, the subscript 1 is not written. It is understood to be 1. Thus, a hydrogen chloride molecule has the formula HCl. What would be the

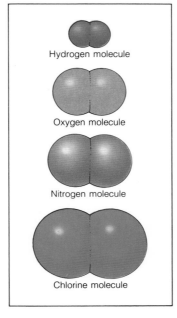

Hydrogen molecule

Oxygen molecule

Nitrogen molecule

Chlorine molecule

Figure 1–16 *Diatomic elements include hydrogen, oxygen, nitrogen, and chlorine. How many valence electrons does an atom of each element have?* ❷

O ■ 23

FACTS AND FIGURES
ANTOINE LAVOISIER

In 1789, Antoine Lavoisier created a list of 33 different substances called elements. Scientists today still acknowledge the majority of these substances as genuine elements.

CONTENT DEVELOPMENT

Point out that in a typical covalent bond, a relatively small number of atoms are involved in the sharing of electrons. In covalent bonds, atoms form separate units.
• **What kind of units were formed by atoms of ionic compounds?** (Large crystal-lattice structures.)

Refer students to the hydrogen atoms of Figure 1–13. Explain that if two covalently bonded atoms are the same, the substance is considered to be an element. Refer students to the atoms pictured in Figure 1–14. Explain that if two covalently bonded atoms are different, such as the hydrogen atom and the chlorine atom in this diagram, the substance is a compound.

GUIDED PRACTICE

Skills Development

Skill: Making diagrams

Provide students with the following electron-dot diagrams: Na, F, K, Cl, H, and I.

Have students draw the electron-dot diagram for the compounds NaF, KCl, and HI and determine the type of bonding that forms between each pair of atoms.

REINFORCEMENT/RETEACHING

▶ *Activity Book*

Students who need practice on the concept of electron bonding should be provided with the Chapter 1 activity called Electron-Dot Diagram. In this activity students will use electron-dot diagrams to show how different atoms can join together to form covalent compounds.

Figure 1–17 *Diamonds are network solids. Network solids contain bonds that are difficult to break. This accounts for the extreme hardness of diamonds. Extremely strong glues are also examples of network solids.*

formula for a molecule that has 1 carbon (C) atom and 4 chlorine (Cl) atoms? ❶

Covalently bonded solids tend to have low melting points. Some covalent substances, however, do not have low melting points. They have rather high melting points. This is because molecules of these substances are very large. The molecules are large because the atoms involved continue to bond to one another. These substances are called **network solids**. Carbon in the form of graphite is an example of a network solid. So too is silicon dioxide, the main ingredient in sand. Certain glues also form networks of atoms whose bonds are difficult to break. This accounts for the holding properties of such glues.

Polyatomic Ions

Certain ions are made of covalently bonded atoms that tend to stay together as if they were a single atom. A group of covalently bonded atoms that acts like a single atom when combining with other atoms is called a **polyatomic ion**. Although the bonds within the polyatomic ion are covalent, the polyatomic ion usually forms ionic bonds with other atoms.

Figure 1–18 is a list of some of the more common polyatomic ions, and Figure 1–19 on page 26 shows the atomic structure of several polyatomic ions. Some of these ions may sound familiar to you. For example, the polyatomic ion hydrogen carbonate (HCO_3^{1-}) bonded to sodium produces sodium hydrogen carbonate ($NaHCO_3$), better known as baking soda. Magnesium hydroxide ($MgOH_2$) is milk of magnesia. And ammonium nitrate (NH_4NO_3—two polyatomic ions bonded together) is an important fertilizer.

POLYATOMIC IONS

Name	Formula
ammonium	NH_4^{1+}
acetate	$C_2H_3O_2^{1-}$
chlorate	ClO_3^{1-}
hydrogen carbonate	HCO_3^{1-}
hydroxide	OH^{1-}
nitrate	NO_3^{1-}
nitrite	NO_2^{1-}
carbonate	CO_3^{2-}
sulfate	SO_4^{2-}
sulfite	SO_3^{2-}
phosphate	PO_4^{3-}

Figure 1–18 *The name and formula of some common polyatomic ions are shown here. What is a polyatomic ion? Which polyatomic ion has a positive charge?* ❷

CONNECTIONS

Diagonals by Design ①

Diamonds by Design ❶

What do a razor blade, a rocket engine, and swamp gas have in common? Hardly anything, you might think. But today, researchers are working to provide an answer. Scientists involved in this endeavor are trying to find a way to coat various objects with a thin film of synthetic diamond. (Synthetic diamonds are made in the laboratory.)

Diamonds are extremely hard and resistant to wear. By placing a diamond coating on a razor blade, the blade can be made to last longer and stay sharper. Diamond coatings on rocket engines and cutting tools will increase their resistance to wear. The list of uses for diamond coatings goes on and on.

How do scientists go about making a synthetic diamond coating? They start with swamp gas, which is called methane. Methane is a chemical substance made of 1 carbon atom linked to 4 hydrogen atoms. The first step in the process is to "strip away" the hydrogen atoms from the carbon atom. When this happens, the carbon atoms from thousands of methane molecules are left behind.

Diamonds are made of carbon atoms. By carefully controlling conditions in the laboratory, scientists can make the carbon atoms link together to form synthetic diamonds. As they link together, they are deposited on the object to be coated. The synthetic diamonds are almost pure.

With continued research in this field of *chemical technology*, scientists feel certain that in any form, diamonds will continue to be extremely valuable!

Synthetic diamonds similar to these shown mixed with natural diamonds can be used to coat various objects, including silicon wafers, thereby making the objects stronger, sharper, and more durable.

O ■ 25

REINFORCEMENT/RETEACHING

Have students select the formulas that represent covalent molecules from the following list:

a. CCl_4
b. CO_2
c. Na_2CO_3
d. KCl

(Both a and b represent covalent molecules.)

Have students determine the number of nitrogen atoms in each of the following nitrogen-containing compounds:

a. $(NH_4)_2CO_3$
b. N_2O
c. HNO_3
d. $Mg(NO_3)_2$

(Choice a contains 2 nitrogen atoms, b contains 2, c contains 1, and d contains 2.)

1-4 Metallic Bonds

MULTICULTURAL OPPORTUNITY 1-4

During the seventeenth and eighteenth centuries, Central and South America were leading sites for metallurgy and mining research. Have students investigate some of this work. They may wish to research the life and work of such scientists as Juan José D'Elhuyar, Fausto D'Elhuyar, and Andres Manuel Del Rio.

ESL STRATEGY 1-4

Have students tell what type of action takes place in metallic bonding. Then have them list the four characteristics of metals and explain the meaning of each characteristic.

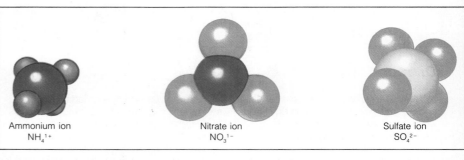

Ammonium ion
NH_4^{1+}

Nitrate ion
NO_3^{1-}

Sulfate ion
SO_4^{2-}

Figure 1-19 *A polyatomic ion is a group of covalently bonded atoms that act like a single atom when combining with other atoms. What kind of bond does a polyatomic ion usually form with another atom?* ❶

1-3 Section Review

1. What is covalent bonding?
2. What is an electron-dot diagram? How is it used to represent a covalent bond?
3. What is a molecule? What does the chemical formula for a molecule tell you?
4. What is a polyatomic ion? Give two examples.

Critical Thinking—*Applying Concepts*
5. What elements and how many atoms of each are represented in the following formulas: Na_2CO_3, $Ca(OH)_2$, $Mg(C_2H_3O_2)_2$, $Ba_3(PO_4)_2$?

Guide for Reading

Focus on this question as you read.

▶ What is a metallic bond?

1-4 Metallic Bonds

You are probably familiar with metals such as copper, silver, gold, iron, tin, and zinc. And perhaps you even know that cadmium, nickel, chromium, and manganese are metals too. But do you know what makes an element a metal? Metals are elements that give up electrons easily.

In a metallic solid, or a solid made entirely of one metal element, only atoms of that particular metal are present. There are no other atoms to accept the electron(s) the metal easily gives up. How, then, do the atoms of a metal bond?

The atoms of metals form **metallic bonds**. In a metallic bond, the outer electrons of the atoms form

1-3 (continued)

INDEPENDENT PRACTICE

▶ *Activity Book*

Students who need practice on the concept of different types of bonds should complete the chapter activity Drawing Chemical Bonds. In this activity students will draw diagrams of the ionic and covalent bonds that are formed in various substances.

GUIDED PRACTICE

▶ *Laboratory Manual*

Skills Development

Skill: *Making comparisons*

At this point you may want to have students complete the Chapter 1 Laboratory Investigation in the *Laboratory Manual* called Comparing Covalent and Ionic Compounds. In this investigation students will determine how the properties of ionic and covalent compounds differ.

INDEPENDENT PRACTICE

Section Review 1-3

1. It is bonding in which electrons are shared.

2. It is a diagram that represents the electron sharing in a covalent bond. In it, the chemical symbol for an element represents the nucleus and inner energy levels. In covalently bonded compounds, it shows how valence electrons are shared, and dots surrounding the symbol represent valence electrons.

3. The smallest particle of a covalently bonded substance that retains all the properties of that substance; the formula contains the symbol of each element involved in the bond and shows the exact number of atoms of each element.

4. A group of covalently bonded atoms that acts like a single atom when combining with other atoms. See Figure 1-18.

5. Two sodium, 1 carbon, 3 oxygen; 1 calcium, 2 oxygen, 2 hydrogen; 1 magnesium, 4 carbon, 6 hydrogen, 4 oxygen; 3 barium, 2 phosphorus, 8 oxygen.

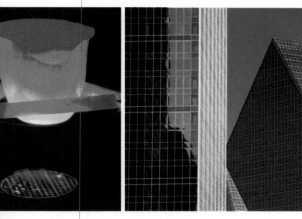

Answers

❶ Ionic. (Relating facts)

Integration

❶ Physical Science: Metals. See *Matter: Building Block of the Universe*, Chapter 5.

a common electron cloud. This common distribution of electrons occurs throughout a metallic crystal. In a sense, the electrons become the property of all the atoms. These electrons are often described as a "sea of electrons." **The positive nuclei of atoms of metals are surrounded by free-moving, or mobile, electrons that are all attracted by the nuclei at the same time.**

The sea of mobile electrons in a metallic crystal accounts for many properties of metals. Metals are malleable, which means they can be hammered into thin sheets without breaking. Metals are also ductile: They can be drawn into thin wire. The flexibility of metals results from the fact that the metal ions can slide by one another and the electrons are free to flow. Yet the attractions between the ions and the electrons hold the metal together even when it is being hammered or drawn into wire.

Figure 1–20 *The atoms of metals form metallic bonds. Metallic bonding accounts for many important properties of metals that make metals very useful. The metal platinum has an extremely high melting point, and so it is used in heat-resistant containers (left). The walls of this building are covered with a thin film of metal that reflects a significant amount of outdoor light (center). Metals are also excellent conductors of electricity. Some metals offer so little resistance to electric current that they can be used as superconductors (right).*

BACKGROUND INFORMATION
METALLIC SOLIDS

Metallic solids are composed of atoms that have a positive core and are surrounded by a "sea of electrons." Typical metallic solids include iron, silver, gold, and copper.

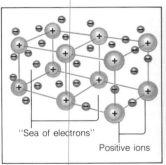

"Sea of electrons"

Positive ions

Figure 1–21 *In a metallic bond, the outer electrons of the metal atoms form a "sea of mobile electrons." Because metals are both malleable and ductile, they can be hammered and drawn into a wide variety of shapes. Here you see various forms of gold.*

O ■ 27

the same location and cause the wire to break apart. Explain how the energy you are adding overcomes the attractive forces.

CONTENT DEVELOPMENT

Because most students have had experiences with metal in its solid phase, it may be easy for them to forget that bonding between atoms within metals involves a situation in which the valence electrons move. Remind students that the diagram in Figure 1–21 uses a "freeze frame" type of picture. In reality, the electrons are continually moving.

● ● ● ● **Integration** ● ● ● ●

Use the introduction of various metals to integrate concepts of metals into your lesson.

REINFORCEMENT/RETEACHING

Monitor students' responses to the Section Review questions. If students appear to have difficulty with any of the questions, review the appropriate material in the section.

CLOSURE

▶ *Review and Reinforcement Guide*

At this point have students complete Section 1–3 in the *Review and Reinforcement Guide.*

TEACHING STRATEGY 1–4

FOCUS/MOTIVATION

Have students look around the classroom and notice all the objects made of metals. If you have samples of pure metals such as tin, aluminum, or gold, distribute them to students for examination.

Display a long piece of copper wire to the class. Bend the wire into various shapes while discussing metallic bonding properties. Bend the wire several times in

1-5 Predicting Types of Bonds

Have students investigate the life and work of Dmitri Mendeleev. Although working in Russia during a time in which that country was not known for great scientific discoveries, Mendeleev profoundly changed our understanding of chemistry by his ability to perceive patterns. He not only organized the world's understanding of elements but was able to successfully predict the properties of then undiscovered elements

ESL STRATEGY 1-5

Ask students these questions:
• What does the oxidation number of an atom describe?
• What do you have to know to determine the oxidation number of any atom?
• What should you know to predict how atoms will combine?
• What must you know in order to be able to state the new formula for a compound?

1-4 (continued)

CONTENT DEVELOPMENT

Make it a point to review with students the different types of bonds they have studied. Have them recall that an ionic bond involves the transfer of electrons and that a covalent bond involves the sharing of electrons. Point out that metallic bonds are formed by the atoms of metals in which the outer electrons form a common electron cloud. Explain that this common electron cloud is often described as a "sea of electrons." It can also be described as a common distribution of outer electrons in which the electrons become the property of all the atoms involved in the bond.

Remind students of the names of some common metals and the properties they typically have.

● ● ● ● **Integration** ● ● ● ●

Use the examples of melting points to integrate heat concepts into your lesson.

Activity Bank

Hot Stuff, p.151

Guide for Reading

Focus on these questions as you read.

▶ *What is oxidation number?*
▶ *What is the relationship between oxidation number and bond type?*

ENRICHMENT

▶ *Activity Book*

Students will be challenged by the Chapter 1 activity in the *Activity Book* called Bond Identification. In this activity students will identify chemical bonds, will diagram electron-dot models, and will write chemical formulas for various atoms and substances.

The ability of the electrons to flow freely makes metals excellent conductors of both heat and electricity. Metallic bonding also accounts for the high melting point of most metals. For example, the melting point of silver is 961.9°C and of gold, 1064.4°C.

1-4 Section Review

1. What is a metallic bond?
2. What is a malleable metal? A ductile metal?
3. How does metallic bonding account for the properties of metals?

Connection—*You and Your World*

4. In terms of bonding, explain why it would be unwise to stir a hot liquid with a silver utensil.

1-5 Predicting Types of Bonds

You have just learned about three different types of bonds formed between atoms of elements: ionic bonds, covalent bonds, and metallic bonds. By knowing some of the properties of an element, is there a way of predicting which type of bond it will form? Fortunately, the answer is yes. And the property most important for predicting bond type is the electron arrangement in the atoms of the element—more specifically, the number of valence electrons.

The placement of the elements involved in bonding in the periodic table often indicates whether the bond will be ionic, covalent, or metallic. Look again at the periodic table on pages 154–155. Elements at the left and in the center of the periodic table are metals. These elements have metallic bonds.

Compounds formed between elements that lose electrons easily and those that gain electrons easily will have ionic bonds. Elements at the left and in the center of the periodic table tend to lose valence electrons easily. These elements are metals. Elements at the right tend to gain electrons readily. These elements are nonmetals. A compound formed between a metal and a nonmetal will thus have ionic bonds.

INDEPENDENT PRACTICE

Section Review 1-4

1. It is a bond formed by the atoms of metals in which the outer electrons of the atoms form a common electron cloud.
2. A malleable metal is able to be hammered into a thin sheet without breaking; a ductile metal is able to be drawn into a thin wire.
3. Metals are both malleable and ductile. This flexibility of metals results from metallic bonding. Metal ions can slide by

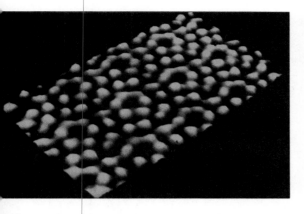

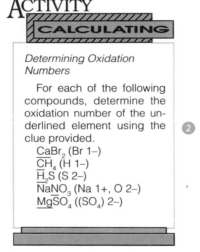

Figure 1–22 *This is the first photograph ever taken of atoms and their bonds. The bright round objects are single atoms. The fuzzy areas between atoms represent bonds.*

Compounds formed between elements that have similar tendencies to gain electrons will have covalent bonds. Bonds between nonmetals, which are at the right of the periodic table, will be covalent. What type of bonding would you expect between magnesium (Mg) and fluorine (F)? Between oxygen (O) and chlorine (Cl)? In a sample of zinc (Zn)? ●

Combining Capacity of Atoms

The number of electrons in the outermost energy level of an atom, the valence electrons, determines how an atom will combine with other atoms. If you know the number of valence electrons in an atom, you can calculate the number of electrons that atom needs to gain, lose, or share when it forms a compound. **The number of electrons an atom gains, loses, or shares when it forms chemical bonds is called its oxidation number**. The **oxidation number** of an atom describes its combining capacity.

An atom of sodium has 1 valence electron. It loses this electron when it combines with another atom. In so doing, it forms an ion with a 1+ charge, Na^{1+}. The oxidation number of sodium is 1+. A magnesium atom has 2 valence electrons, which it will lose when it forms a chemical bond. The magnesium ion is Mg^{2+}. The oxidation number of magnesium is 2+.

An atom of chlorine has 7 valence electrons. It will gain 1 electron when it bonds with another atom. The ion formed will have a 1– charge, Cl^{1-}. The oxidation number of chlorine is 1–. Oxygen has

Activity Bank

The Milky Way, p.152

ACTIVITY
CALCULATING

Determining Oxidation Numbers

For each of the following compounds, determine the oxidation number of the underlined element using the clue provided. ●

$Ca\underline{Br}_2$ (Br 1–)
$\underline{C}H_4$ (H 1–)
$\underline{H}_2S$ (S 2–)
$Na\underline{N}O_3$ (Na 1+, O 2–)
$Mg\underline{S}O_4$ ((SO$_4$) 2–)

O ■ 29

You may want to have students write a short report or give a brief oral synopsis of the book to the class.

Integration: Use this Activity to integrate language arts into your science lesson.

ECOLOGY NOTE
ADVERSE EFFECTS OF ELEMENTS

Using the premise of the Activity, challenge students to use reference materials to find another situation in which an element has had an adverse effect on living things in the United States and then share their findings with the class.

1–5 (continued)

CONTENT DEVELOPMENT

This section integrates all the concepts of the chapter. Emphasize the understanding of the relationship between an element's position on the periodic table and its bonding type and bonding capacity.

GUIDED PRACTICE

Skills Development

Skill: Making comparisons

Some very common substances have more than one oxidation number. To illustrate this point, have students predict the formulas for iron combined with chloride when iron is in the +2 state and in the +3 state. Also have students predict the formulas for copper combined with chloride when copper is in the +1 state and the +2 state. ($FeCl_2$; $FeCl_3$; $CuCl$; $CuCl_2$.) Have students note that tin may also be a +2 or +4.

Figure 1–23 *Some elements have more than one oxidation numb Copper, seen here as pennies, can have oxidation numbers of 1+ and 2+. Mercury also exhibits these same oxidation numbers. Ho* *can you determine the oxidation number of an atom?* ❶

6 valence electrons. How many electrons will it gain What is its oxidation number? ❷

ACTIVITY
READING

Dangerous Food

❶ Sometimes an element can have adverse effects on living things and on the environment. Such unwanted effects are often the result of bonding between the element and another substance. *Minamata,* a book by W. Eugene Smith and Aileen M. Smith, describes the tragic consequences to the Japanese population from the dumping of industrial mercury into Minamata Bay. You may find this story of the poisoning of a bay fascinating and important reading.

Using Oxidation Numbers

You can use the oxidation numbers of atoms to predict how atoms will combine and what the formula for the resulting compound will be. In orde to do this, you must follow one important rule: *The sum of the oxidation numbers of the atoms in a compoun must be zero.*

Sodium has an oxidation number of 1+. Chlorin has an oxidation number of 1−. One atom of sodiur will bond with 1 atom of chlorine to form NaCl. Magnesium has an oxidation number of 2+. When magnesium bonds with chlorine, 1 atom of magnesium must combine with 2 atoms of chlorine, since each chlorine atom has an oxidation number of 1−. In other words, 2 atoms of chlorine are needed to gain the electrons lost by 1 atom of magnesium. Th compound formed, magnesium chloride, contains 2 atoms of chlorine for each atom of magnesium. Its formula is $MgCl_2$. What would be the formula for calcium bromide? For sodium oxide? Remember th rule of oxidation numbers!

• **What would be the formula of the two possible chlorides of tin?** ($SnCl_2$; $SnCl_4$.)
• **The following are formulas of acids containing an atom of chlorine. What is the oxidation number of chlorine in each compound?** $HClO_4$; $HClO_3$; $HClO_2$; $HClO$; HCl? (+7; +5; +3, +1, -1.)

INDEPENDENT PRACTICE

▶ *Activity Book*

Students who need practice on the concept of the combining capacity of atoms should complete the chapter activity

Charting Oxidation Number. In this ac tivity students will identify the structure o given elements and predict the oxidation number of those elements.

ENRICHMENT

▶ *Activity Book*

Students will be challenged by the Chapter 1 activity in the *Activity Book* called Chemical Analogies. In this activity students will identify relationships between various science terms.

PROBLEM ??? Solving

A Little Kitchen Chemistry

While preparing lunch one day in home economics class, Monduane and Juan noticed something unusual. As they added a dash of sugar to the vegetable oil they were using to make salad dressing, the sugar dissolved completely. But when they poured in a small amount of salt, as called for in the recipe they were following, the salt did not dissolve. Curious about their observation, the two chefs decided to repeat the process. They observed the same results. What, they wondered, could account for this difference in behavior? Would other substances show differences in their ability to dissolve in vegetable oil? In water?

Anxious to find the answers to their questions, the two students sought the advice of their teacher. The answer they received was simply this: Like dissolves in like.

Developing and Testing a Hypothesis

1. Help Monduane and Juan with their problem by suggesting a hypothesis.

2. Design an experiment to test your hypothesis. Remember to include a control.

3. What other applications does your hypothesis explain?

1–5 Section Review

1. What is an oxidation number?
2. How can the oxidation number of an atom be determined?
3. How is the oxidation number related to bond type?
4. What rule of oxidation numbers must be followed in writing chemical formulas?

Critical Thinking—*Making Predictions*
5. Predict the type of bond for each combination: Ca–Br, C–Cl, Ag–Ag, K–OH, SO_4^{2-}.

PROBLEM SOLVING
A LITTLE KITCHEN CHEMISTRY

This activity gives students an opportunity to relate concepts from their studies of the chapter to a real-life situation. Hypotheses suggested by students will vary according to experiment designs and applications, depending on the student. Check to ensure that a given design will not only test the hypothesis of each student and include a control but also adhere as much as possible to the scientific method.

bonding. From this information, bond type can be predicted, as well as the formula for the resulting compounds.
4. The sum of the oxidation numbers of the atoms in a compound must be zero.
5. Ionic; covalent; metallic; ionic; covalent.

REINFORCEMENT/RETEACHING

Monitor students' responses to the Section Review questions. If students appear to have difficulty with any of the questions, review the appropriate material in the section.

CLOSURE

▶ *Review and Reinforcement Guide*
At this point have students complete Section 1–5 in the *Review and Reinforcement Guide.*

CONTENT DEVELOPMENT

Have students recall the picture of salt crystals at the beginning of this chapter. The chapter introduction asked these questions: How and why do these elements (sodium and chlorine) combine to form salt? What is the process by which hundreds of thousands of other substances are formed? Pose these questions to students and have volunteers offer answers. Then have students recall the concepts they learned about in this chapter and pose questions that could have been asked on the introductory page, but were not.

INDEPENDENT PRACTICE

Section Review 1–5
1. The number of electrons an atom gains, loses, or shares when it forms a chemical bond.
2. By the number of valence electrons in the atom.
3. The oxidation number of an atom describes its combining capacity—how many electrons it will gain, lose, or share when

Laboratory Investigation

PROPERTIES OF IONIC AND COVALENT COMPOUNDS

BEFORE THE LAB

Gather all materials and equipment at least one day prior to the investigation. Be sure to have enough materials and equipment to meet your class needs, assuming six students per group.

PRE-LAB DISCUSSION

Have students read the complete laboratory procedure.

• **What is the purpose of this investigation?** (To determine if covalent compounds have properties different from those of ionic compounds.)

• **In an ionic compound, what is the regular, repeating arrangement of ions called?** (A crystal lattice.)

• **What is the combination of atoms formed by a covalent bond called?** (A molecule.)

Ask students to develop a hypothesis describing what they predict they will observe in this investigation and have them record the hypothesis in their notes.

SAFETY TIPS

Use safety precautions when working with Bunsen burners and when working with electricity.

Laboratory Investigation

Properties of Ionic and Covalent Compounds

Problem

Do covalent compounds have different properties from ionic compounds?

Materials (per group)

safety goggles	distilled water
salt	(200 mL)
4 medium-sized	2 100-mL
test tubes	beakers
glass-marking	stirring rod
pencil	3 connecting
test-tube tongs	wires
Bunsen burner	light bulb socket
timer	light bulb
sugar	dry-cell battery
vegetable oil	

Procedure

1. Place a small sample of salt in a test tube. Label the test tube. Place an equal amount of sugar in another test tube. Label that test tube.

2. Using tongs, heat the test tube of salt over the flame of the Bunsen burner. **CAUTION:** *Observe all safety precautions when using a Bunsen burner.* Determine how long it takes for the salt to melt. Immediately stop heating when melting begins. Record the time.

3. Repeat step 2 using the sugar.

4. Half fill a test tube with vegetable oil. Place a small sample of salt in the test tube. Shake the test tube gently for about 10 seconds. Observe the results.

5. Repeat step 4 using the sugar.

6. Pour 50 mL of distilled water into a 100-mL beaker. Add some salt and stir until it is dissolved. To another 100-mL beaker add some sugar and stir until dissolved.

7. Using the beaker of salt water, set up a circuit as shown. **CAUTION:** *Exercise care when using electricity.* Observe the results. Repeat the procedure using the beaker of sugar water.

Observations

1. Does the salt or the sugar take a longer time to melt?

2. Does the salt dissolve in the vegetable oil? Does the sugar?

3. Which compound is a better conductor of electricity?

Analysis and Conclusions

1. Which substance do you think has a higher melting point? Explain.

2. Explain why one compound is a better conductor of electricity than the other.

3. How do the properties of each type of compound relate to their bonding?

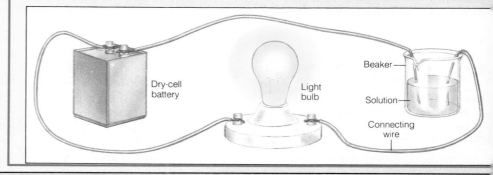

Dry-cell battery   Light bulb   Beaker   Solution   Connecting wire

---

TEACHING STRATEGIES

1. Circulate throughout the classroom during the investigation to ensure that students have set up the investigation properly, are following the specified laboratory procedure, and are observing all safety precautions.

2. You may discover that groups may not be able to melt salt with the Bunsen burner. If this occurs, have groups stop trying to melt the salt after a reasonable period of time. The salt will still be able to be used to describe the characteristics of ionic compounds.

3. At the end of the investigation, ask students to compare their initial hypothesis to their actual observations.

DISCOVERY STRATEGIES

Discuss how the investigation relates to the chapter by asking open questions similar to the following:

• **Vegetable oil is a covalent compound. If "like dissolves in like," predict the type of bonding in salt if salt does not dissolve in**

Study Guide

Summarizing Key Concepts

1–1 What Is Chemical Bonding?

▲ All matter is made of tiny particles called atoms.

▲ Chemical bonding is the combining of elements to form new substances.

▲ The atom consists of a positively charged nucleus containing protons and neutrons, and energy levels containing electrons.

▲ Bonding involves the electrons in the outermost energy level, the valence electrons.

1–2 Ionic Bonds

▲ Ionic bonding involves a transfer of electrons and a formation of ions.

▲ Ionization energy is the amount of energy needed to remove an electron from a neutral atom. Electron affinity is the tendency of an atom to attract electrons.

▲ The placement of ions in an ionic compound results in a crystal lattice.

1–3 Covalent Bonds

▲ Covalent bonding involves a sharing of electrons.

▲ A molecule is the smallest unit of a covalently bonded substance.

▲ Network solids are substances whose molecules are very large because the atoms in the substance continue to bond to one another.

▲ A polyatomic ion is a group of covalently bonded atoms that acts like a single atom when it combines with other atoms.

1–4 Metallic Bonds

▲ The basis of metallic bonding is the sea of mobile electrons that surrounds the nuclei and is simultaneously attracted by them.

1–5 Predicting Types of Bonds

▲ The oxidation number, or combining capacity, of an atom refers to the number of electrons the atom gains, loses, or shares when it forms chemical bonds.

▲ The oxidation number of any atom can be determined by knowing the number of electrons in its outermost energy level.

Reviewing Key Terms

Define each term in a complete sentence.

1–1 What Is Chemical Bonding?
atom
chemical bonding
valence electron

1–2 Ionic Bonds
ionic bonding
ion
ionization
ionization energy
electron affinity
crystal lattice

1–3 Covalent Bonds
covalent bonding
electron-dot diagram
diatomic element
molecule
network solid
polyatomic ion

1–4 Metallic Bonds
metallic bond

1–5 Predicting Types of Bonds
oxidation number

O ■ 33

ANALYSIS AND CONCLUSIONS

1. Salt. It took longer for the salt to melt, if it did in fact melt.

2. The crystal lattice structure of an ionic compound lends itself to conducting electricity better than the more separate units of a covalent bond do.

3. Covalent compounds typically have low melting points, dissolve in covalent solvents, and do not conduct an electric current. Ionic compounds typically have high melting points, do not dissolve in covalent solvents, and conduct electricity when dissolved in water.

GOING FURTHER: ENRICHMENT

Part 1

Provide students with unknown compounds and have them conduct tests similar to those in this investigation to classify the unknown compounds according to bond type.

Part 2

Provide students with samples of metallic and covalently bonded substances. Ask students to design an experiment that compares the samples for malleability, ductility, and melting point.

Chapter Review

ALTERNATIVE ASSESSMENT

The *Prentice Hall Science* Program includes a variety of testing components and methodologies. Aside from the Chapter Review questions, you may opt to use the Chapter Test or the Computer Test Bank Test in your *Test Book* for assessment of important facts and concepts. In addition, Performance-Based Tests are included in your *Test Book*. These Performance-Based Tests are designed to test science process skills, rather than factual content recall. Since they are not content dependent, Performance-Based Tests can be distributed after students complete a chapter or after they complete the entire textbook.

---

vegetable oil. (Because salt does not dissolve in vegetable oil, salt must be an ionic compound.)

• **Vegetable oil is a covalent compound. If "like dissolves in like," predict the type of bonding in sugar if sugar dissolves in vegetable oil.** (Because sugar dissolves in vegetable oil, sugar must be a covalent compound.)

• **The salt used in this activity has a crystal structure. What would you predict about the shape of these crystals?** (The crystals are likely to have flat surfaces and very distinct, right-anglelike corners.)

OBSERVATIONS

1. Salt.
2. No; yes.
3. Salt is the better conductor because it forms ions when dissolved in water.

CONTENT REVIEW

Multiple Choice

1. c
2. d
3. c
4. c
5. c
6. c
7. a
8. a
9. a
10. b

True or False

1. F, valence
2. F, Ionic
3. T
4. F, oxidation number
5. F, Network
6. F, an ion
7. T

Concept Mapping

Row 1: New Substances
Row 2: Atoms; Energy Levels;
    Electrons
Row 3: Protons; Neutrons

CONCEPT MASTERY

1. In an ionic bond, electrons are transferred from one atom to another. In a covalent bond, electrons are shared between atoms. In a metallic bond, the outer electrons of an atom form a common electron cloud.

2. The elements have their outermost shells filled to capacity with valence electrons and, as a result, are quite stable or unreactive and not likely to form chemical bonds.

3. Metals are both malleable and ductile. This flexibility of metals results from metallic bonding. Metal ions can slide by one another, and the electrons can flow freely while the attractions between the ions and the electrons hold the metal together. The ability of the electrons to flow freely also accounts for the high electric conductivity of metals. Metals typically have high melting points due to the strength of their metallic bond. A great deal of heat is needed to overcome the attraction between the ions and electrons in the bond.

4. The oxidation number of an atom describes its combining capacity—how many electrons it will gain, lose, or share when bonding. From this information, bond type can be predicted as well as the for-

Chapter Review

Content Review

Multiple Choice

Choose the letter of the answer that best completes each statement.

1. Chemical bonding is the combining of elements to form new
   a. atoms.   c. substances.
   b. energy levels.   d. electrons.
2. The center of an atom is called the
   a. electron.   c. octet.
   b. energy level.   d. nucleus.
3. The maximum number of electrons in the second energy level is
   a. 1.   b. 2.   c. 8.   d. 18.
4. Bonding that involves a transfer of electrons is called
   a. metallic.   c. ionic.
   b. covalent.   d. network.
5. Bonding that involves sharing of electrons within a molecule is called
   a. metallic.   c. covalent.
   b. ionic.   d. crystal.
6. The combination of atoms formed by covalent bonds is called a(an)
   a. element.   c. molecule.
   b. ion.   d. crystal.

7. Atoms that readily lose electrons have
   a. low ionization energy and low electron affinity.
   b. high ionization energy and low electron affinity.
   c. low ionization energy and high electron affinity.
   d. high ionization energy and high electron affinity.
8. An example of a polyatomic ion is
   a. SO_4^{2-} b. Ca^{2-} c. $NaCl$. d. O_2.
9. A sea of electrons is the basis of bonding in
   a. metals.
   b. ionic substances.
   c. nonmetals.
   d. covalent substances.
10. Bonding between atoms on the left and right sides of the periodic table tends to be
   a. covalent.   c. metallic.
   b. ionic.   d. impossible.

True or False

If the statement is true, write "true." If it is false, change the underlined word or words to make the statement true.

1. Electrons in the outermost energy level are called <u>oxidation</u> electrons.
2. <u>Covalent</u> bonds form crystals.
3. The tendency of an atom to attract electrons is called <u>electron affinity</u>.
4. The combining capacity of an atom is described by its <u>crystal lattice</u>.
5. <u>Malleable</u> solids are substances whose molecules are very large.
6. A charged atom is called a <u>molecule</u>.
7. Bromine is a <u>diatomic</u> element.

Concept Mapping

Complete the following concept map for Section 1–1. Refer to pages 06–07 to construct a concept map for the entire chapter.

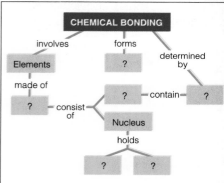

mula for the resulting compounds.

5a. A crystal lattice is a regular, repeating arrangement of ions in a compound formed by ionic bonds. This arrangement gives the compound great stability and accounts for its typically high melting point. The arrangement also accounts for the particular shape of the crystals of the compound. **b.** A network solid is a substance formed by covalent bonds. The molecules of the substance are large, and its bonds are usually difficult to break. A network solid usually has a high melting

point. **c.** A covalently bonded solid is formed from weak covalent bonds. Only a small amount of energy is needed to break these bonds, and so the melting point of this solid tends to be low.

6a. Ionization energy is the energy needed to remove an electron from an atom to form an ion. Electron affinity is the tendency of an atom to attract electrons. **b.** Atoms with high electron affinity attract electrons easily, and atoms with low electron ionization energy lose electrons easily. As a result, when atoms with high elec-

Concept Mastery

Discuss each of the following in a brief paragraph.

1. List the three types of chemical bonds and explain the differences among them.
2. Explain why the elements of Family 18 do not tend to form chemical bonds.
3. What are four properties of metals? How does the bonding in metals account for these properties?
4. How can you use the oxidation number of an atom to predict how it will bond?
5. Define the following structures that result from chemical bonds. Give one physical property of each.
   a. crystal lattice
   b. network solid
   c. covalently bonded solid
6. a. What is the difference between ionization energy and electron affinity?
   b. Why do atoms of high electron affinity tend to form ionic compounds with atoms of low ionization energy?
7. How can the periodic table be used to predict bond types?
8. Explain why the elements in Family 1 and in Family 17 are highly reactive.
9. How does chemical bonding account for the fact that although there are only 109 different elements, there are hundreds of thousands of different substances.
10. Describe the formation of a positive ion. Of a negative ion.

Critical Thinking and Problem Solving

Use the skills you have developed in this chapter to answer each of the following.

1. **Making predictions** Predict the type of bond formed by each pair of atoms. Explain your answers.
   a. Mg and Cl   c. I and I
   b. Na and Na   d. Li and I
2. **Making diagrams** Draw the electron configuration for a Period 2 atom from each of the following families of the periodic table: Family 1, 2, 13, 14, 15, 16, 17, 18.
3. **Identifying patterns** Use the periodic table to predict the ion that each atom will form when bonding.
   a. sulfur (S)   d. astatine (At)
   b. rubidium (Rb)   e. sodium (Na)
   c. argon (Ar)   f. aluminum (Al)
4. **Applying facts** Draw an electron-dot diagram for the following molecules and explain why they are stable: F_2, NF_3.
5. **Making predictions** Use the periodic table to predict the formulas for the compounds formed by each of the following pairs of atoms.
   a. K and S   c. Ba and S
   b. Li and F   d. Mg and N

6. **Applying concepts** Explain why a sodium ion is smaller than a sodium atom.
7. **Identifying patterns** As you can see in the accompanying photograph, a sheet of aluminum metal is being cut. The ability of a metal to be cut into sections without shattering is called sectility. How can you account for the sectility of aluminum?

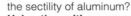

8. **Using the writing process** Pretend you are an experienced electron about to give an orientation lecture for "incoming freshmen electrons." Prepare your speech, being sure to include important details about placement in the atom, ionic bonding, covalent bonding, and metallic bonding.

O ■ 35

---

tron affinity are combined with atoms with low ionization energy, the atoms with high electron affinity gain electrons from the atoms with low ionization energy.

7. The periodic table can be used to determine the number of valence electrons in the outermost level of an atom, and this electron arrangement determines whether or not the atom will form chemical bonds and the type of bonds it will form.

8. These elements have oxidation numbers of plus one (+1) or minus one (−1) and, as a result, are quite reactive and

likely to form chemical bonds.

9. Atoms can combine in many different ways, which form many different substances.

10. A neutral atom that loses one or more electrons becomes a positive ion. A neutral atom that gains one or more electrons becomes a negative ion.

CRITICAL THINKING AND PROBLEM SOLVING

1a. Ionic because Mg is a metal and Cl is a nonmetal; **b.** Metallic because Na is a metal; **c.** Covalent because I is a nonmetal; **d.** Ionic because Li is a metal and I is a nonmetal.
2. Li: 2,1; Be: 2,2; B: 2,3; C: 2,4; N: 2,5; O: 2,6; F: 2,7; Ne: 2,8.
3a. S^{2-}; **b.** Rb^{1+}; **c.** Ar does not usually form ions; **d.** At^{1-}; **e.** Na^{1+}; **f.** Al^{3+}.   **4.** Check students' drawings. Both molecules are stable because the atoms that they are composed of have complete outer energy levels.
5a. K_2S; **b.** LiF; **c.** BaS; **d.** Mg_3N_2.
6. A sodium ion has two electron energy levels; a sodium atom has three electron energy levels.
7. Answers may vary. Students might suggest that the ability of the electrons to flow freely allows a cut without shattering.
8. Speeches will vary but should detail the placement of an electron in an atom and the role of an electron in ionic, covalent, and metallic bonding.

KEEPING A PORTFOLIO

You might want to assign some of the Concept Mastery and Critical Thinking and Problem Solving questions as homework and have students include their responses to unassigned questions in their portfolio. Students should be encouraged to include both the question and the answer in their portfolio.

ISSUES IN SCIENCE

The following issue can be used as a springboard for discussion or given as a writing assignment:

New medicines frequently appear on the market. Use reference materials or visit a local chemist or pharmaceutical company to find out how chemists experiment to develop new substances. Also find out about the testing procedures for new drugs. Are the procedures too strict? Why or why not?

O ■ 35

Chapter 2 CHEMICAL REACTIONS

SECTION	HANDS-ON ACTIVITIES

2–1 Nature of Chemical Reactions
pages O38–O41
 	
Multicultural Opportunity 2–1, p. O38
 	
ESL Strategy 2–1, p. O38
	Student Edition
	  ACTIVITY (Discovering): Flashes and Masses, p. O39
	Activity Book
	  CHAPTER DISCOVERY: Kitchen Chemistry, p. O37
	Teacher Edition
	  How Does a Flashbulb Work? p. O36d

2–2 Chemical Equations pages O42–O46 Multicultural Opportunity 2–2, p. O42 ESL Strategy 2–2, p. O42	**Student Edition** ACTIVITY BANK: Pocketful of Posies, p. O153

2–3 Types of Chemical Reactions
pages O46–O50
 	
Multicultural Opportunity 2–3, p. O46
 	
ESL Strategy 2–3, p. O46
	Student Edition
	  ACTIVITY (Discovering): Preventing a Chemical Reaction, p. O47
	  ACTIVITY (Doing): The Disappearing Coin, p. O48
	  ACTIVITY (Discovering): Double-Replacement Reaction, p. O49
	  ACTIVITY BANK: Popcorn Hop, p. O154
	Laboratory Manual
	  Chemical Synthesis, p. O15
	  Chemical Decomposition, p. O19
	  Single-Replacement Reactions, p. O23
	  Double-Replacement Reactions, p. O27
	Activity Book
	  ACTIVITY BANK: Cartoon Chemistry, p. O153

2–4 Energy of Chemical Reactions
pages O50–O52
 	
Multicultural Opportunity 2–4, p. O50
 	
ESL Strategy 2–4, p. O50
	Student Edition
	  ACTIVITY BANK: Toasting to Good Health, p. O155
	Teacher Edition
	  Zinc Replaces Hydrogen, p. O36d

2–5 Rates of Chemical Reactions
pages O53–O57
 	
Multicultural Opportunity 2–5, p. O53
 	
ESL Strategy 2–5, p. O53
	Student Edition
	  ACTIVITY (Discovering): Rate of Reaction, p. O54
	  ACTIVITY (Discovering): Temperature and Reaction Rate, p. O55
	  LABORATORY INVESTIGATION: Determining Reaction Rate, p. O58
	Laboratory Manual
	  Comparing Reaction Rate and Catalysts, p. O31

Chapter Review pages O58–O61	

OUTSIDE TEACHER RESOURCES
Books
Denn, Morton M. *Stability of Reaction and Transport Processes*, Prentice-Hall.

Skinner, Gordon B. *Introduction to Chemical Kinetics*, Academic Press.

Sykes, Peter. *The Search for Organic Reaction Pathways*, Halstead.

Audiovisuals
Balancing Equations, courseware, Prentice-Hall

Chemical Change All About Us, 16-mm film, Coronet

OTHER ACTIVITIES	MEDIA AND TECHNOLOGY
Student Edition ACTIVITY (Reading): The Loss of the Hindenburg, p. O41 **Activity Book** ACTIVITY: Conservation of Mass, p. O41 **Review and Reinforcement Guide** Section 2–1, p. O15	◈ **Courseware** Physical and Chemical Properties (Supplemental) 🎧 **English/Spanish Audiotapes** Section 2–1
Student Edition ACTIVITY (Calculating): A Balancing Act, p. O45 **Activity Book** ACTIVITY: Balancing Equations, p. O43 ACTIVITY: Identifying and Balancing Chemical Equations, p. O49 ACTIVITY: Completing Equations, p. O47 ACTIVITY: Chemical Formulas and Equations, p. O51 **Review and Reinforcement Guide** Section 2–2, p. O17	🎧 **English/Spanish Audiotapes** Section 2–2
Activity Book ACTIVITY: Types of Chemical Reactions, p. O46 **Review and Reinforcement Guide** Section 2–3, p. O19	🎧 **English/Spanish Audiotapes** Section 2–3
Student Edition ACTIVITY (Thinking): Kitchen Chemistry, p. O52 **Review and Reinforcement Guide** Section 2–4, p. O21	🎧 **English/Spanish Audiotapes** Section 2–4
Activity Book ACTIVITY: Testing Reaction Rate Factors, p. O39 **Review and Reinforcement Guide** Section 2–5, p. O23	🎧 **English/Spanish Audiotapes** Section 2–5
Test Book Chapter Test, p. O31 Performance-Based Tests, p. O117	◈ **Test Book** Computer Test Bank Test, p. O37

*All materials in the Chapter Planning Guide Grid are available as part of the Prentice Hall Science Learning System.

Chemical Change and Temperature, 16-mm film, BFA
Chemical Changes, filmstrip with cassette, Society for Visual Education
Chemical Changes in Everyday Life, video, Encyclopaedia Britannica Education
Chemical Reaction, filmstrip, Encyclopaedia Britannica Education
Chemical Reactions, Queue
Combustion—An Introduction to Chemical Change, 16-mm film, BFA
Dynamic Equilibrium, courseware, Prentice-Hall

Chapter 2 CHEMICAL REACTIONS

CHAPTER OVERVIEW

A chemical reaction is any process in which the chemical properties of the original substances disappear as new substances with new properties form.

There are four major types of chemical reactions: synthesis, decomposition, single replacement, and double replacement. In all chemical reactions, energy is involved. Chemical reactions can release energy or absorb energy. A reaction that releases energy is called an exothermic reaction. A reaction that absorbs energy is called an endothermic reaction.

The chemical reactions of three major groups of compounds—acids, bases, and salts—account for the formation of many familiar substances. These compounds have individual properties that enable them to react in a definite way.

Chemical reactions result in chemical changes. Chemical reactions are written as equations that consist of reactants and products. Mass and energy must be conserved in any chemical reaction; that is, the mass and energy of the reactants must be equal to the mass and energy of the products. A chemical reaction follows the law of conservation of mass and energy.

2–1 NATURE OF CHEMICAL REACTIONS

THEMATIC FOCUS

The purpose of this section is to introduce students to the characteristics of chemical reactions. They will learn that a chemical reaction always produces a change in the properties and a change in energy of the substances involved in the reaction. Students will be introduced to the terms *reactants* and *products* and will come to understand that a chemical reaction involves the breaking and forming of chemical bonds.

The themes that can be focused on in this section are patterns of change and scale and structure.

*Patterns of change: A chemical reaction changes original substances into new substances with different chemical and physical properties, including a different amount of energy.

*Scale and structure: The ability of a substance to undergo a chemical reaction depends on the number of valence electrons in its atoms. Collisions among the particles of the reactants determine the rate of a chemical reaction.

PERFORMANCE OBJECTIVES 2–1

1. **Describe the characteristics of chemical reactions.**
2. **Explain that a chemical reaction is accompanied by a change in properties and a change in energy of the substances involved in the reaction.**
3. **Define the terms *reactant* and *product*.**
4. **Explain how a substance's capacity to react is related to the arrangement of electrons in the outermost energy level of its atoms.**

SCIENCE TERMS 2–1
chemical reaction p. O38
reactant p. O39
product p. O39

2–2 CHEMICAL EQUATIONS

THEMATIC FOCUS

The purpose of this section is to show students how chemical equations can be used to represent chemical reactions. They will learn that elements are represented by symbols and that compounds are represented by formulas. By the end of this section, students will come to understand that a chemical equation illustrates the law of conservation of mass.

The themes that can be focused on in this section are systems and interactions and stability.

*Systems and interactions: Chemical equations can be used to represent the changes that occur during a chemical reaction.

*Stability: Mass is always conserved during a chemical reaction.

PERFORMANCE OBJECTIVES 2–2

1. **Discuss how chemical equations are used to describe chemical reactions.**
2. **Explain how a chemical equation illustrates the law of conservation of mass.**
3. **Balance chemical equations.**

SCIENCE TERMS 2–2
chemical equation p. O42

2–3 TYPES OF CHEMICAL REACTIONS

THEMATIC FOCUS

The purpose of this section is to introduce students to four general types of chemical reactions: synthesis, decomposition, single replacement, and double replacement. Students will also learn that the opposite of a synthesis reaction is a decomposition reaction.

The themes that can be focused on in this section are patterns of change and unity and diversity.

*Patterns of change: In a chemical reaction, the bonds between atoms are broken, and new bonds are formed.

Unity and diversity: All chemical reactions share certain characteristics, such as changing reactants into products. There are, however, different types of chemical reactions: synthesis reactions, decomposition reactions, single-replacement reactions, and double-replacement reactions.

PERFORMANCE OBJECTIVES 2–3

1. Describe and cite examples of a synthesis reaction.
2. Describe and cite examples of a decomposition reaction.
3. Describe and cite examples of a single-replacement reaction and a double-replacement reaction.

SCIENCE TERMS 2–3

synthesis reaction p. O47
decomposition reaction p. O47
single-replacement reaction p. O48
double-replacement reaction p. O49

2–4 ENERGY OF CHEMICAL REACTIONS

THEMATIC FOCUS

The purpose of this section is to show students that chemical reactions can be classified according to energy changes. They will learn that an exothermic reaction is one in which energy is released and that an endothermic reaction is one in which energy is absorbed.

The themes that can be focused on in this section are energy and systems and interactions.

*Energy: The energy of the reactants changes during a chemical reaction. A chemical reaction can give off energy (exothermic) or absorb energy (endothermic.) The reactants must be able to reach the activation energy before the products can be formed.

*Systems and interactions: All chemical reactions require activation energy. The effect of this energy can be shown with an energy diagram.

PERFORMANCE OBJECTIVES 2–4

1. Distinguish between exothermic and endothermic reactions.
2. Define and discuss the term *activation energy*.

SCIENCE TERMS 2–4

exothermic reaction p. O50
endothermic reaction p. O51
activation energy p. O52

2–5 RATES OF CHEMICAL REACTIONS

THEMATIC FOCUS

The purpose of this section is to introduce students to four factors that influence rates of chemical reactions. These factors are concentration of reactants, surface area of solid reactants, temperature, and the presence of a catalyst.

The themes that can be focused on in this section are systems and interactions and energy.

*Systems and interactions: Increasing the temperature, concentration, or surface area of the reactants usually increases reaction rate.

*Energy: A catalyst can increase reaction rate by decreasing the required activation energy.

PERFORMANCE OBJECTIVES 2–5

1. Relate the collision theory to factors affecting rates of chemical reactions.
2. List four factors that affect the rate of a chemical reaction.
3. Explain the role of a catalyst in a chemical reaction.

SCIENCE TERMS 2–5

kinetics p. O53
reaction rate p. O53
collision theory p. O53
catalyst p. O55

Discovery *Learning*

TEACHER DEMONSTRATIONS
MODELING
How Does a Flashbulb Work?

For this demonstration, you will need several flashbulbs and a camera with a flash attachment. (It is not necessary to have film in the camera.) Students should have pencil and paper to record their observations throughout the demonstration.

Display the flashbulbs and allow students to examine the bulbs carefully.
• **What do you see inside the bulb?** (A small coil of shiny, gray metal.)

• **What do you think is occupying the rest of the space inside the bulb?** (Answers may vary; some students may say "nothing" or "air." The bulb actually contains oxygen.)

Place a flashbulb in the camera and set it off.
• **What did you see?** (A flash of light.)

Set off several more flashbulbs and allow students to observe carefully the used bulbs.
• **What do you see inside the bulb?** (White powder.)
• **How does the appearance of the bulb differ from its appearance before it was set off?** (Before, the inside was clear, and the coil of metal was plainly visible; now the entire bulb is clouded with white powder.)

Have students try to describe what they think happened when the bulb went off. Most will realize that a chemical reaction occurred. Point out that they will discover the nature of this reaction as they read this chapter.

Zinc Replaces Hydrogen

For this demonstration you will need a piece of zinc and a beaker of hydrochloric acid. Display the zinc; then carefully place it in the beaker. Ask students to notice the bubbles of gas that are released. Write the equation for this reaction on the chalkboard.

$$Zn + 2HCl \rightarrow ZnCl_2 + H_2$$

• **What were the bubbles of gas that you saw?** (Hydrogen.)
• **Where did the hydrogen come from?** (From the hydrochloric acid.)
• **According to the equation, how did the zinc change in this reaction?** (It joined with the chlorine to produce the salt, zinc chloride.)

CHAPTER 2
Chemical Reactions

INTEGRATING SCIENCE

This physical science chapter provides you with numerous opportunities to integrate other areas of science, as well as other disciplines, into your curriculum. Blue numbered annotations on the student page and integration notes on the teacher wraparound pages alert you to areas of possible integration.

In this chapter you can integrate physical science and chemical changes (p. 38), photography (p. 39), physical science and heat (p. 40), social studies (p. 41), language arts (pp. 41, 50), physical science and chemical formulas (p. 42), earth science and volcanoes (p. 43), physical science and conservation of mass (p. 43), mathematics (p. 45), physical science and rusting (p. 47), food science (p. 48), life science and digestion (p. 49), physical science and energy changes (p. 51), earth science and ecology (p. 56), physical science and petrochemicals (p. 56), and earth science and atmosphere (p. 57).

SCIENCE, TECHNOLOGY, AND SOCIETY/COOPERATIVE LEARNING

Acid precipitation is produced by a series of chemical reactions in which sulfur and nitrogen oxides combine with water and oxygen in the atmosphere to produce sulfuric and nitric acid. Emissions of sulfur and nitric oxides primarily come from coal-burning factories and power plants and automobile exhaust. Acid precipitation is destroying freshwater aquatic and plant life and reducing plant growth and reproduction. It is also considered a threat to human health.

Acid precipitation has become a major political issue between the United States and Canada. Pollution from midwestern factories and power plants in the United States is carried by prevailing winds into Canada, where the acid precipitates and damages Canada's economically important timber industry and fishing-related tourism. The political and economic issue of who should pay to clean up acid-rain damage has been hotly debated between the governments of the United States and Canada since the early 1980s.

INTRODUCING CHAPTER 2

DISCOVERY LEARNING

▶ *Activity Book*

Begin your introduction to this chapter by using the Chapter 2 Discovery Activity from the *Activity Book*. Using this activity, students will use salt, baking soda, and starch to investigate factors that affect the speed of chemical reactions.

USING THE TEXTBOOK

Bronze is an alloy of copper and tin. Exposure to air and moisture had caused a patina, or thin film of corrosion, to form on the Statue of Liberty. The chemical reaction for this process, which involves copper, is $Cu + CO_2 + H_2O \rightarrow Cu(OH)_2CO_3$.

Direct students' attention to the photograph of the Statue of Liberty.

Encourage any students who have visited the statue to share their impressions of it. If they were there before the restora-

Chemical Reactions

Fireworks flash brilliantly in the night sky over the dark waters of the harbor. It is Independence Day. Amidst the wonderful celebration stands a very special lady. She towers above the waters, the torch in her upraised hand reaching high into the sky. She is a symbol of freedom, justice, and the brotherhood of people of all nations. Her name is Liberty.

She has stood there for over a century. But the passage of time had not been especially kind to her. The bronze of her outer structure, once bright and gleaming, had turned a dull gray-green. And the structure that supports her had begun to weaken. What had caused these changes? The answer has to do with the chemistry of atoms.

This chemistry, which damaged the Statue of Liberty, had also made possible the glorious restoration of this Lady in the Harbor. And the colorful fireworks lighting up the sky are products of the chemistry of atoms.

Chemical changes take place all the time, not just on the Fourth of July. And they take place everywhere, not just in New York Harbor. In this chapter you will learn about the nature of these chemical changes, many of which shape the world around you.

Journal *Activity*

You and Your World All around you, matter undergoes permanent changes. When bread is baked, for example, the dough changes forever. You can never turn the bread back into dough. A burned log can never be unburned. In your journal, describe several examples of substances that undergo changes and turn into new substances.

n July 4, 1986, fireworks lit up the sky in New York Harbor as the nation celebrated e one-hundredth birthday of the Statue of Liberty.

O ■ 37

Cooperative learning: Using a pre-assigned lab groups or randomly selected teams, have groups complete one of the following assignments:

• Have groups prepare a grid (alternatives along the top of the chart and consequences along the left margin) identifying the economic, political, social, and environmental consequences of the alternatives listed below:

A. No government regulations on industry regarding sulfuric and nitric oxide emissions.

B. Air pollution standards are rewritten to significantly reduce acid-rain-causing emissions.

C. Groups prepare their own alternative.

Upon completion of their grids, have groups identify the alternative they think our government should follow in dealing with the issue of acid rain.

• Have groups apply the five themes of geography listed below to the problem of acid rain.

Region: What region(s) of North America are most affected by acid rain?

Human-Environment Interaction: How are humans affected by acid rain?

Location: Where are acid-rain pollutants produced?

Movement: What carries pollutants from their source to the area of acid-rain damage?

Place: Describe the effects of acid rain on the lakes and forests of northeastern Canada.

See Cooperative Learning in the *Teacher's Desk Reference.*

JOURNAL ACTIVITY

You may want to use the Journal Activity as the basis of a class discussion. Explain that in their journal entries, students should try to describe chemical changes. Students should be instructed to keep their Journal Activity in their portfolio.

---

ion, it is possible that they were aware of he statue's need for repair.

What do you think caused the changes hat made it necessary to repair the stat-ue? (Exposure to air and water.)

• **Do you think these changes were phys-cal changes? Why not?** (No. The proper-ies of the metal changed—not only did ts color change, but the metal in the sup-port structure lost some of its strength.)

• **What kind of changes were these?** Chemical changes.)

• **What must be done to reverse or repair a chemical change?** (There must be another chemical change.)

• **Do any of you know what bronze is made of?** (It is an alloy of copper and tin.)

Emphasize to students that an alloy is a mixture—actually a solid solution—of two or more metals. Point out that in the corrosion of the Statue of Liberty, the copper reacted with water and the atmosphere.

2-1 Nature of Chemical Reactions

Guide for Reading
Focus on this question as you read.
▶ What are the characteristics of a chemical reaction?

Figure 2-1 *A burnt match has undergone a chemical reaction. So has rusted metal. Chemical reactions are also responsible for producing the vibrant colors of autumn leaves.*

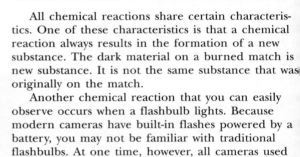

2-1 Nature of Chemical Reactions

You have probably never given much thought to an ordinary book of matches. But stop for a minute and consider the fact that a single match in a book of matches can remain unchanged indefinitely. Yet if someone strikes that match, it bursts into a brilliant flame. And when that flame goes out, the appearance of the match will have changed forever. It can never be lighted again. The match has undergone a **chemical reaction.** What does this mean? **A chemical reaction is a process in which the physical and chemical properties of the original substances change as new substances with different physical and chemical properties are formed.** The burning of gasoline, the rusting of iron, and the baking of bread are all examples of chemical reactions.

Characteristics of Chemical Reactions

All chemical reactions share certain characteristics. One of these characteristics is that a chemical reaction always results in the formation of a new substance. The dark material on a burned match is new substance. It is not the same substance that was originally on the match.

Another chemical reaction that you can easily observe occurs when a flashbulb lights. Because modern cameras have built-in flashes powered by a battery, you may not be familiar with traditional flashbulbs. At one time, however, all cameras used flashbulbs similar to those shown in Figure 2–2 to provide the light necessary to take a photograph. Such flashbulbs can be used only once. You will now find out why.

Inside a flashbulb is a small coil of shiny gray metal. This metal is magnesium. The bulb is filled with the invisible gas oxygen. When the flashbulb is set off, the magnesium combines with the oxygen in a chemical reaction. During the reaction, energy is released in the form of light, and a fine white powder is produced. You can see this powder on the inside of the bulb. The powder is magnesium oxide, a compound with physical and chemical properties unlike those of the elements that were originally present—magnesium and oxygen. During the chemical reaction, the original substances are changed into a new substance. Now you can understand why traditional flashbulbs can be used only once.

The substances present before the change and the substances formed by the change are the two kinds of substances involved in a chemical reaction. A substance that enters into a chemical reaction is called a **reactant** (ree-AK-tuhnt). A substance that is produced by a chemical reaction is called a **product**. So a general description of a chemical reaction can be stated as reactants changing into products. In the example of the flashbulb, what are the reactants? The product? ❶

In addition to changes in chemical and physical properties, chemical reactions always involve a

Figure 2–2 *Inside a flashbulb, oxygen surrounds a thin coil of magnesium. When the flashbulb is set off, a chemical reaction takes place in which magnesium combines with oxygen to form magnesium oxide. How can you tell a chemical reaction has occurred?* ❷

O ■ 39

ECOLOGY NOTE
FOOD PRESERVATIVES

There are increasing concerns among scientists and the general public over the use of chemical additives in food. Food manufacturers add these substances to food products for a variety of reasons.

Particularly worrisome are the nitrates and nitrites used as preservatives in meat. Over the years, people have come to prefer the cured flavor these additives give to meat, so the substances are now added both to slow down spoilage and to give taste.

Nitrates, however, are not harmless. They react with the hemoglobin in human blood to form methemoglobin, a compound that cannot carry oxygen. And nitrites can react with the amines present in foods to form nitrosamines, many of which have been shown to be carcinogenic in test animals.

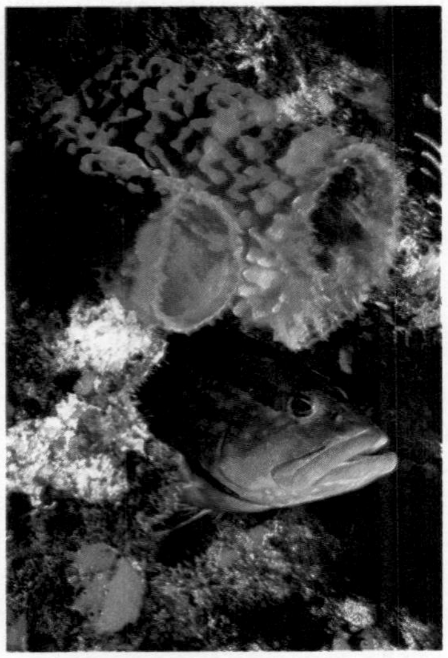

Figure 2–3 *The chemical reactions that occur within sea coral regulate the amount of carbon dioxide in the ocean. Why do you think this is important?* ①

change in energy. Energy is either absorbed or released during a chemical reaction. For example, he energy is absorbed when sugar changes into carame When gasoline burns, heat energy is released. Later in this chapter you will learn more about the energ changes that accompany chemical reactions.

Capacity to React

In order for a chemical reaction to occur, the reactants must have the ability to combine with oth er substances to form products. What accounts for the ability of different substances to undergo certai chemical reactions? In order to answer this question you must think back to what you learned about atoms and bonding.

Atoms contain electrons, or negatively charged particles. Electrons are located in energy levels surrounding the nucleus, or center of the atom. The electrons in the outermost energy level are called the valence electrons. It is the valence electrons tha are involved in chemical bonding. An atom forms chemical bonds with other atoms in order to complete its outermost energy level. As you learned in Chapter 1, having a complete outermost energy lev is the most stable condition for an atom. An atom will try to fill its outermost energy level by gaining or losing electrons, or by sharing electrons. A chem cal bond formed by the gain or loss of electrons is an ionic bond. A chemical bond formed by the sha ing of electrons is a covalent bond.

The arrangement of electrons in an atom determines the ease with which the atom will form chem cal bonds. An atom whose outermost energy level i

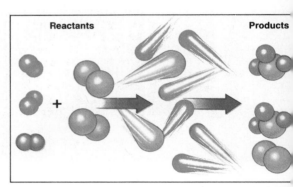

Reactants   **Products**

Figure 2–4 *During a chemical reaction, bonds between atoms of reactants are broken, atoms are rearranged, and new bonds in products are formed.*

2–1 (continued)

CONTENT DEVELOPMENT

Lead into a discussion of valence electrons by reviewing the properties of noble gases.
• **Why are the noble gases so unreactive?** (The outermost energy level of electrons in a noble gas is complete.)
• **How many electrons are needed to make up a complete outermost energy level?** (Eight, unless the first energy level of the atom is also the outermost level—then only 2 electrons are needed.)

● ● ● ● **Integration** ● ● ● ●

Use the discussion of gasoline to integrate concepts of heat into your lesson.

GUIDED PRACTICE

Skills Development

Skills: Making observations, making predictions

Show students diagrams of lithium and fluorine atoms.
• **How many electrons are in the outermost energy level of lithium?** (1).

• **How many electrons are in the outermost energy level of fluorine?** (7).
• **How would you predict that lithium would combine with another atom to fill its outermost energy level?** (It would lose 1 electron.)
• **Can you make a prediction about what would happen if lithium and fluorine were combined chemically with each other?** (They would form an ionic bond in which lithium would give up an electron to fluorine.)

ull will not bond with other atoms. But an atom with an incomplete outermost energy level will bond readily. The ease with which an atom will form chemical bonds is known as the bonding capacity of an atom. The bonding capacity of an atom determines its ability to undergo chemical reactions. And the ability to undergo chemical reactions is an important chemical property.

During a chemical reaction, atoms can form molecules, molecules can break apart to form atoms, or molecules can react with other molecules. In any case, new substances are produced as existing bonds are broken, atoms are rearranged, and new bonds are formed.

2–1 Section Review

1. What is a chemical reaction?
2. What is a reactant? A product?
3. What is the relationship between the arrangement of electrons in an atom and the atom's chemical properties?

Critical Thinking—*Applying Concepts*

4. Cooking an egg until it is hard-boiled involves a chemical reaction. Cutting a piece of paper into a hundred little pieces does not involve a chemical reaction. Explain the difference between the two processes.

ACTIVITY READING

The Loss of the Hindenburg

The air was thick with excitement on May 6, 1937, as the German airship (blimp) *Hindenburg* headed toward its landing site in Lakehurst, New Jersey. The airship, held aloft by 210,000 cubic meters of hydrogen gas, was completing another transatlantic voyage. But before it could moor, the excitement turned to horror as the ship burst into flames. To learn about the *Hindenburg* disaster, read *Horror Overhead* by Richard A. Boning.

O ■ 41

ACTIVITY READING

THE LOSS OF THE *HINDENBURG*

Skill: Reading comprehension

Explain that a *dirigible* is any lighter-than-air aircraft that is engine-driven and steerable. The *Hindenburg* was, at the time of its launch, the largest dirigible ever constructed. During its first year of operation, the *Hindenburg* completed ten round trips across the North Atlantic between Germany and New Jersey without mishap. Then, during a landing in New Jersey, the disaster described in the feature occurred.

The generally accepted explanation for the cause of the tragedy was that atmospheric electricity from a recent storm ignited hydrogen leaking from the ship.

Point out that the *Hindenburg* disaster received widespread publicity when it occurred. Most subsequent airships were inflated with safer, nonexplosive helium.

Integration: Use this Activity to integrate language arts into your science lesson.

new substances with different physical and chemical properties are formed.
2. A reactant is a substance that enters into a chemical reaction. A product is a substance that is produced by a chemical reaction.
3. The arrangement of electrons in an atom determines the bonding capacity of an atom. The bonding capacity of an atom determines its chemical properties, or its ability to undergo chemical reactions.
4. Cooking an egg involves a chemical reaction that produces new substances. The physical act of cutting paper does not produce a new substance.

REINFORCEMENT/RETEACHING

Review students' responses to the Section Review questions. Reteach any material that is still unclear, based on students' responses.

CLOSURE

▶ *Review and Reinforcement Guide*

Have students complete Section 2–1 in the *Review and Reinforcement Guide.*

CONTENT DEVELOPMENT

Explain that an atom forms chemical bonds with other atoms in order to complete its outermost energy level and that in these bonds an atom can lose, gain, or share electrons.

● ● ● ● **Integration** ● ● ● ●

Use the photographs of the *Hindenburg* and the Statue of Liberty to integrate social studies concepts into your lesson.

INDEPENDENT PRACTICE

▶ *Activity Book*

Students will be introduced to the concept of conservation of mass in the Chapter 2 activity called Conservation of Mass.

INDEPENDENT PRACTICE

Section Review 2–1

1. A chemical reaction is a process in which the physical and chemical properties of the original substances change as

2-2 Chemical Equations

Have students research the life and work of Dr. Percy Julian (1899–1975). Julian was not only a leading African-American chemist but was also a leader in encouraging African-American students to enter careers in science. His primary area of research was the synthesis of artificial pharmaceuticals. He developed an inexpensive synthetic means for producing cortisone, which is used in the treatment of arthritis.

ESL STRATEGY 2-2

Ask students to complete the following paragraph:

In a chemical reaction, electrons are _____, _____, or shared. However, the total _____ does not change. This is known as the law of _____ of _____. (**Answers:** gained, lost, mass, conservation, mass.)

Ask for volunteers to answer each of the following questions:
1. What is used by scientists to describe a chemical reaction? (A chemical equation.)
2. How is this type of description different from one in which words are used? (Answers will vary. Possible answers include that the equation is shorter or more precise.)
3. On which side of the equation do you place the reactants? The products? (The left; the right.)
4. What must always be the same on both sides of a chemical equation? (The total number of atoms of each element.)

Guide for Reading

Focus on these questions as you read.
▶ What is the law of conservation of mass?
▶ How are chemical equations balanced?

CAREERS

Food Chemist

Today, most methods of food processing involve chemical reactions. **Food chemists** use their knowledge of chemistry to develop these food processing methods.

Some food chemists develop new foods or new flavors. Others develop improved packaging and storage methods for foods.

If you are interested in a career as a food chemist, write to the Institute of Food Technology, Suite 300, 221 North LaSalle Street, Chicago, IL 60601.

2-2 Chemical Equations

It is important to be able to describe the details of a chemical reaction—how the reactants changed into the products. This involves indicating all the individual atoms involved in the reaction. One way of doing this is to use words. But describing a chemical reaction with words can be awkward. Many atoms may be involved, and the changes may be complicated.

For example, consider the flashbulb reaction described earlier. A word equation for this reaction would be: Magnesium combines with oxygen to form magnesium oxide and give off energy in the form of light. You could shorten this sentence by saying: Magnesium and oxygen form magnesium oxide and light energy.

Chemists have developed a more convenient way to represent a chemical reaction. Using symbols to represent elements and formulas to represent compounds, a chemical reaction can be described by a **chemical equation**. A chemical equation is an expression in which symbols and formulas are used to represent a chemical reaction.

In order to write a chemical equation, you must first write the correct chemical symbols or formulas for the reactants and products. Then you need to show that certain substances combine. This is done with the use of a + sign, which replaces the word and. Between the reactants and the products, you need to draw an arrow to show that the reactants have changed into the products. The arrow, which is read "yields," takes the place of an equal sign. It also shows the direction of the chemical change. The chemical equation for the flashbulb reaction can now be written:

$$Mg + O_2 \longrightarrow MgO + Energy$$
$$\text{Magnesium} + \text{Oxygen} \longrightarrow \text{Magnesium oxide} + \text{Energy}$$

Conservation of Mass

Chemists have long known that atoms can be neither created nor destroyed during a chemical reaction. In other words, the number of atoms of each

---

TEACHING STRATEGY 2-2

FOCUS/MOTIVATION

• **Ask a student volunteer to go to the chalkboard and write in words the chemical reaction that takes place when a flashbulb is set off.** (The student's choice of words may vary but should convey that magnesium reacts with oxygen to produce magnesium oxide and energy in the form of light.)
• **Ask a second volunteer to go to the chalkboard and write in words the chemical re-**

action that produces rust. (Iron combines with oxygen to produce iron oxide.)

CONTENT DEVELOPMENT

Explain that chemical symbols and equations provide a standard form of expression that scientists all over the world understand. Use the descriptions of reactions that students wrote on the chalkboard during the Focus/Motivation activity to lead into a discussion of the need for a more convenient way to represent chemical reactions.

You may find it helpful when teaching this section to have a periodic table on display so that students can readily note the symbols for various elements.

● ● ● ● **Integration** ● ● ● ●

Use the discussion of chemical equations to integrate concepts of chemical formulas in your lesson.

REINFORCEMENT/RETEACHING

At this point you may wish to review the information included in the period-

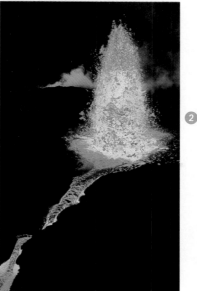

element must be the same before and after the chemical reaction (that is, the number of atoms remains the same on both sides of the arrow in a chemical equation). The changes that occur during any chemical reaction involve only the rearrangement of atoms, not their production or destruction.

Every atom has a particular mass. Because the number of atoms of each element remains the same, mass can never change in a chemical reaction. The total mass of the reactants must equal the total mass of the products. No mass is lost or gained. **The observation that mass remains constant in a chemical** ③ **reaction is known as the law of conservation of mass.**

Balancing Chemical Equations

The law of conservation of mass must be taken into account when writing a chemical equation for a chemical reaction. A chemical equation must show that atoms are neither created nor destroyed. The number of atoms of each element must be the same on both sides of the equation.

Figure 2–6 *Whether a chemical reaction is involved in the use of pesticides to protect crops, the violent eruption of a volcano, or the awesome explosion of a hydrogen bomb, one characteristic is always the same: Mass is never lost. What law describes this observation?* ①

O ■ 43

sented as O_2, which indicates a molecule of oxygen containing 2 oxygen atoms.)

Write on the chalkboard "Mg + O_2." Then draw an arrow and explain that the arrow means "yields" or "produces."
• **What is the product in this reaction?** (Magnesium oxide.)

Complete the equation by writing the formula for magnesium oxide, MgO.
• **What does this formula tell you about the composition of a molecule of magnesium oxide?** (It contains 1 atom of magnesium and 1 atom of oxygen.)

● ● ● ● **Integration** ● ● ● ●

Use the photograph of the volcanic eruption to integrate volcanoes into your lesson.

Use the discussion of the total mass of chemical reactants to integrate conservation of mass into your lesson.

INDEPENDENT PRACTICE
▶ *Activity Book*

Students who need practice with the concepts of this section should be provided with the Chapter 2 activity called Balancing Equations.

ENRICHMENT
▶ *Activity Book*

Students who have mastered the concepts in this section will be challenged by the Chapter 2 activity called Identifying and Balancing Chemical Equations.

ic table of the elements.

Reproduce on the chalkboard the enlarged element key for carbon (C). Ask a student volunteer to go to the chalkboard and reproduce the element key for chlorine (Cl). Use the enlarged models to point out the location of the element's name, symbol, atomic number, and atomic mass. Emphasize to students that the first letter of a chemical symbol is always capitalized, whereas the second letter is not.

CONTENT DEVELOPMENT

Refer to the description of the magnesium and oxygen reaction that is on the chalkboard.
• **What two elements are the reactants in this reaction?** (Magnesium and oxygen.)
• **What are the symbols for these elements?** (Mg and O.)

Write "Mg + O" on the chalkboard.
• **Does the O look strange to you as a symbol for oxygen?** (It should; remind students that free oxygen is always repre-

ANNOTATION KEY

Answers

1 No, it is not balanced, and the law of conservation of mass has not been observed. (Applying concepts)

2 Place the coefficient 2 in front of Mg on the left side and in front of MgO on the right side. (Making inferences)

3 The law of conservation of mass. (Applying concepts)

Integration

1 Mathematics

2–2 (continued)

CONTENT DEVELOPMENT

Work through on the chalkboard the balancing of the magnesium oxide equation as shown in the textbook.

$$2Mg + O_2 \rightarrow 2MgO$$

Use the finished equation to point out the difference between the 2s used as coefficients and the 2 used as a subscript within the formula for oxygen.

REINFORCEMENT/RETEACHING

Remind students that a chemical equation is just a statement of chemical change using symbols. Use sodium chloride as an example.

• **What familiar substance is created when a sodium atom and a chlorine atom combine?** (Sodium chloride, or table salt.)
• **What is the chemical symbol for sodium? For chlorine?** (Na and Cl.)
• **How would you use symbols to represent the compound sodium chloride?** (NaCl.)

CONTENT DEVELOPMENT

Point out that when balancing chemical equations, the subscripts are never changed. The "trick" is to manipulate the coefficients so that there are equal numbers of atoms on each side of the equation.

As a second example, use the formation of sulfur dioxide. Remind students that when any substance, including sulfur, burns in air, it combines with oxygen to produce an oxide. The skeleton equation for this would be

An equation in which the number of atoms of each element is the same on both sides of the equation is called a balanced chemical equation. Let's go back to the chemical equation for the flashbulb reaction:

$$Mg + O_2 \longrightarrow MgO + Energy$$

Is this a balanced chemical equation? Is the law of conservation of mass observed? 1

How many magnesium atoms do you count on the left side of the equation? You should count 1. And on the right side? You should count 1. Now try the same thing for oxygen. There are 2 oxygen atoms on the left but only 1 on the right. This cannot be correct because atoms can be neither created nor destroyed during a chemical reaction. How, then, can you make the number of atoms of each element the same on both sides of the equation? 2

One thing you cannot do to balance an equation is change a subscript. As you should remember from Chapter 1, a subscript is a small number placed to the lower right of a symbol. Changing the subscript would mean changing the substance. You can, however, change the number of atoms or molecules of each substance involved in the chemical reaction. You can do this by placing a number known as a coefficient (koh-uh-FIHSH-uhnt) in front of the appropriate symbols and formulas. Suppose the coefficient 3 is placed in front of a molecule of oxygen. It would be written as $3O_2$, and it would mean that there are 6 atoms of oxygen (3 molecules of 2 atoms each).

Now let's return to the flashbulb equation. To balance this equation, you must represent more than 1 atom of oxygen on the product side of the equation. If you place a coefficient of 2 in front of the formula for magnesium oxide, you will have 2 molecules of MgO. So you will have 2 atoms of oxygen. But you will also have 2 atoms of magnesium on the product side and only 1 atom of magnesium on the reactant side. So you must add a coefficient of 2 to the magnesium on the reactant side of the equation. There—the equation is balanced:

$$2 Mg + O_2 \longrightarrow 2 MgO + Energy$$

Figure 2–7 *These are the steps to follow in balancing a chemical equation. What law must a chemical equation obey?* 3

BALANCING EQUATIONS

1. Write a chemical equation with correct symbols and formulas.

$$H_2 + O_2 \rightarrow H_2O$$

2. Count the number of atoms of each element on each side of the arrow.

3. Balance atoms by using coefficients.

$$2H_2 + O_2 \rightarrow 2H_2O$$

4. Check your work by counting atoms of each element.

$$S + O_2 \rightarrow SO_2$$

In this example there is no need to change the coefficients because there are already 3 atoms on each side.

As a final example, use the formation of copper dioxide. The skeleton equation is

$$Cu + O_2 \rightarrow CuO$$

Here, you can make the oxygen atoms equal by adding 2 to the right side:

$$Cu + O_2 \rightarrow 2CuO$$

Then, balance the copper part of the equation by changing the left side so that it has 2 copper atoms.

$$2Cu + O_2 \rightarrow 2CuO$$

INDEPENDENT PRACTICE

▶ *Activity Book*

Students who have not mastered the concepts in this section should complete the Chapter 2 activity called Completing Equations for concept reinforcement.

If you count atoms again, you will find 2 magnesium atoms on each side of the equation, as well as oxygen atoms. The equation can be read: 2 atoms of magnesium combine with 1 molecule of oxygen to yield 2 molecules of magnesium oxide. Notice that when no coefficient is written, such as in front of the molecule of oxygen, the number is understood to be 1. Remember that to balance a chemical equation, you can change coefficients but never symbols or formulas.

Chemical equations are actually easy to write and balance. Follow the rules in Figure 2–7 and those listed here.

. Write a word equation and then a chemical equation for the reaction. Make sure the symbols and formulas for reactants and products are correct.
. Count the number of atoms of each element on each side of the arrow. If the numbers are the same, the equation is balanced.
. If the number of atoms of each element is not the same on both sides of the arrow, you must balance the equation by using coefficients. Put a coefficient in front of a symbol or formula so that the number of atoms of that substance is the same on both sides of the arrow. Continue this procedure until you have balanced all the atoms.
. Check your work by counting the atoms of each element to make sure they are the same on both sides of the equation.

2–2 Section Review

1. What is a chemical equation?
2. State the law of conservation of mass.
3. Why must a chemical equation be balanced?
4. Write a balanced chemical equation for the reaction between sodium and oxygen to form sodium oxide, Na_2O.
5. Why can't you change symbols, formulas, or subscripts in order to balance a chemical equation?

Connection—*Mathematics*
6. How are chemical equations related to equations found in mathematics?

ACTIVITY CALCULATING

A Balancing Act

Rewrite each of the following equations on a sheet of paper and balance each.

$BaCl_2 + H_2SO_4 \rightarrow BaSO_4 + HCl$ ❶

$P + O_2 \rightarrow P_4O_{10}$

$KClO_3 \rightarrow KCl + O_2$

$C_3H_8 + O_2 \rightarrow CO_2 + H_2O$

$Cu + AgNO_3 \rightarrow Cu(NO_3)_2 + Ag$

ACTIVITY CALCULATING
A BALANCING ACT

Skill: Computational, making predictions, applying concepts

This activity reinforces students' ability to balance chemical equations. Students should balance the equations as follows:
1. $BaCl_2 + H_2SO_4 = BaSO_4 + 2HCl$
2. $4P + 5O_2 = P_4O_{10}$
3. $2KClO_3 = 2KCl + 3O_2$
4. $C_3H_8 + 5O_2 = 3CO_2 + 4H_2O$
5. $Cu + 2AgNO_3 = Cu(NO_3)_2 + 2Ag$

Integration: Use this Activity to integrate mathematics into your science lesson.

stroyed in a chemical reaction. Thus, the number of atoms on both sides of a chemical equation for each element must be equal, or balanced.
4. $4Na + O_2 = 2Na_2O$
5. Symbols, subscripts, and formulas stand for products or reactants in a chemical equation. If you change them, then the chemical reaction being described is also changed.
6. In both chemical and mathematical equations, the quantity on the right side of the equation must equal the quantity on the left side. An arrow is used instead of an equal sign in a chemical equation.

REINFORCEMENT/RETEACHING

Review students' responses to the Section Review questions. Reteach any material that is still unclear, based on students responses.

CLOSURE

▶ *Review and Reinforcement Guide*
Have students complete Section 2–2 in the *Review and Reinforcement Guide*.

REINFORCEMENT/RETEACHING

Have students write word equations for each of the following chemical equations. Make sure they name each element and compound correctly.
1. $2H_2 + O_2 \rightarrow 2H_2O$
2. $4Fe + 3O_2 \rightarrow 2Fe_2O_3$
3. $2Na + Cl_2 \rightarrow 2NaCl$

ENRICHMENT

▶ *Activity Book*
Students who have mastered the concepts in this section will be challenged by the Chapter 2 activity called Chemical Formulas and Equations.

INDEPENDENT PRACTICE
Section Review 2–2
1. A chemical equation is an expression in which symbols and formulas are used to represent a chemical reaction.
2. The law of conservation of mass says that the mass of reactants equals the mass of the products in a chemical reaction.
3. Atoms can never be created or de-

The Problem Solving feature provides students with an opportunity to apply the concepts they have learned in this chapter. Students should be able to explain that the silverware has tarnished because the silver has combined with oxygen in the air. The resulting silver oxide can be removed by rubbing the tarnished silverware with the rubber band. If students check the label on a container of silver polish, they may discover that the cleaner contains a sulfur compound such as sulfuric acid.

2-3 Types of Chemical Reactions

MULTICULTURAL OPPORTUNITY 2-3

As a way to help students appreciate the field of chemistry, invite a chemist or chemical engineer to speak to them. Often, local businesses and industries will be helpful in identifying speakers.

ESL STRATEGY 2-3

When discussing the four types of chemical reactions, point out that *synthesis* and *decomposition* are antonyms (words that are opposite in meaning). Ask students to define both words.

Have students match the following four types of chemical reactions with their scientific descriptions.
1. $C \rightarrow A + B$
2. $A + BX \rightarrow AX + B$
3. $AX + BY \rightarrow AY + BX$
4. $A + B \rightarrow C$
(**Answers: 1.** decomposition, **2.** single-replacement, **3.** double-replacement, **4.** synthesis.)

PROBLEM ? ? ? Solving

The Silverware Mystery

The Smith family was about to sit down to a special dinner in honor of Mr. Smith's birthday when the telephone rang. The voice on the other end of the telephone informed Mrs. Smith that the family had won a contest. The prize was a fabulous trip that began immediately.

The family excitedly packed suitcases and prepared to leave the house. They wrapped their dinner and put it into the freezer. But they decided to leave the table as it was—beautifully arranged with flowers, china dishes, and silverware. They locked the doors and went on their way!

Two weeks later, after a wonderful trip, the Smiths returned. As they entered the dining room, they noticed something odd. The once shiny silver forks, knives, and spoons were now a dull, blackish-gray color. What could have happened? Could the silverware be returned to its original condition? Being a helpful neighbor, you decide to assist the Smiths in solving the silverware mystery in the following way.

■ You tell them that it's all a matter of chemistry.

■ You give them a rubber band, which contains sulfur, to experiment with.

■ You ask them to think about what the air and the rubber band have in common.

Reaching Conclusions

1. What has happened to the Smiths' silverware?

2. How can it be returned to its original condition?

Guide for Reading

Focus on this question as you read.

▶ *What are the four types of chemical reactions?*

2-3 Types of Chemical Reactions

There are billions of different chemical reactions. In some reactions, elements combine to form compounds. In other reactions, compounds break down into elements. And in still other reactions, one element replaces another.

Chemists have identified four general types of reactions: synthesis, decomposition, single replacement, and double replacement. In each type of reaction, atoms are being rearranged and substances are being changed in a specific way.

TEACHING STRATEGY 2-3

FOCUS/MOTIVATION

Bring in a bottle of seltzer water. Pour some seltzer into a glass and let the class observe it.

• **Do you know what chemical reaction is taking place in this seltzer water?** (Carbonic acid is breaking down to form carbon dioxide and water.)

• **What evidence do you see of this reaction?** (Bubbles of gas are moving up through the water.)

Write on the chalkboard the equation for this reaction.

$$H_2CO_3 \rightarrow H_2O + CO_2$$

CONTENT DEVELOPMENT

Refer to the equation you just wrote in the Focus/Motivation section.

• **How is this equation similar to other equations you have seen so far?** (Reactants are changing into products, as shown by formulas and an arrow.)

Synthesis Reaction

In a **synthesis** (SIHN-thuh-sihs) **reaction**, two or more simple substances combine to form a new, more complex substance. So that you can easily identify synthesis reactions, it may be helpful for you to remember the form these reactions always take:

$$A + B \longrightarrow C$$

For example, the reaction between sodium and chlorine to form sodium chloride is a synthesis reaction:

$$2Na + Cl_2 \longrightarrow 2NaCl$$

Sodium + Chlorine $\longrightarrow$ Sodium chloride

Reactions involving the corrosion of metals are synthesis reactions. The rusting of iron involves the chemical combination of iron with oxygen to form iron oxide. Reactions in which a substance burns in oxygen are often synthesis reactions. Think back to the reaction that occurs in a flashbulb.

Decomposition Reaction

In a **decomposition reaction**, a complex substance breaks down into two or more simpler substances. Decomposition reactions are the reverse of synthesis reactions. Decomposition reactions take the form:

$$C \longrightarrow A + B$$

ACTIVITY
DISCOVERING

Preventing a Chemical Reaction

1. Obtain two large nails. Paint one nail and let it dry. Do not paint the other nail.

2. Pour a little water into a jar or beaker.

3. Stand both nails in the container of water. Cover the container and let it stand for several days. Compare the appearance of the nails.

Describe what happens to each nail. Give a reason for your observations. Write a word equation for any reaction that has occurred.

■ What effect does paint have on the rusting process of a nail?

Figure 2–8 *A reaction in which a substance burns in oxygen is a synthesis reaction. Why do you think smoking is not permitted in areas where oxygen is being administered?* ①

O ■ 47

ANNOTATION KEY

Answers

① Because of the risk of fire or explosion. (Making inferences)

Integration

① Physical Science: Rusting. See *Matter: Building Block of the Universe*, Chapter 2.

ACTIVITY
DISCOVERING
PREVENTING A CHEMICAL REACTION

Discovery Learning

Skills: Applying concepts, making observations, making comparisons

Materials: 2 large iron nails, paint, jar or beaker

Students should discover that the paint slows down or prevents rusting. The unpainted nail will begin to rust as the oxygen in the water combines with the iron to form iron oxide.

Activity Bank

Cartoon Chemistry, Activity Book, p. O153. This activity can be used for ESL and/or Cooperative Learning.

• **Do you notice anything different about this equation compared with others you have seen?** (So far, the equations used in the textbook have consisted of two substances combining to form a product. In this equation a single substance is breaking down to form two products.)

● ● ● ● **Integration** ● ● ● ●

Use the synthesis reactions in metallic corrosion to integrate concepts of rusting into your lesson.

GUIDED PRACTICE

▶ *Laboratory Manual*

Skills Development

Skills: Making observations, applying concepts, making inferences

At this point you may want to have students complete the Chapter 2 Laboratory Investigation in the *Laboratory Manual* called Chemical Decomposition. In the investigation students will decompose water

by electrolysis and will study the elements that are produced.

GUIDED PRACTICE

▶ *Laboratory Manual*

Skills Development

Skills: Making observations, applying concepts, making inferences

At this point you may want to have students complete the Chapter 2 Laboratory Investigation in the *Laboratory Manual* called Chemical Synthesis. In the investigation students will synthesize copper oxide by heating copper metal in air.

ACTIVITY
DOING
THE DISAPPEARING COIN

Skills: Applying concepts, making observations, making inferences

Materials: aluminum foil, glass of water, copper coin

In this activity students will be able to observe the interactions of two different metals in water. Because aluminum is a more active metal than copper, it can replace the copper in the coin.

HISTORICAL NOTE
SODA WATER

Soda water was originally made by combining sulfuric acid with marble dust in a double-replacement reaction that produces carbonic acid and a salt. It has been reported that St. Patrick's Cathedral in New York City supplied marble chips that produced more than 94 million liters of soda water.

2–3 (continued)

CONTENT DEVELOPMENT

Write on the chalkboard the examples given in the textbook of single-replacement and double-replacement reactions. Use arrows to diagram the process of one element or group of atoms replacing another.

● ● ● ● **Integration** ● ● ● ●

Use the photograph of the carbonated beverage to integrate food science concepts into your lesson.

REINFORCEMENT/RETEACHING

The following analogies may help students understand the processes of single replacement and double replacement. For single replacement, draw the analogy of two people playing tennis with a third person watching on the sidelines. When one of the players gets tired, the person on the sidelines comes in and takes his or her place. For double replacement, draw

Figure 2–9 *Carbonic acid gives liquids their "fizz." Carbonic acid, however, quickly decomposes into water and carbon dioxide gas. The reaction that occurs when explosives are ignited is also a decomposition reaction.*

Activity Bank

Popcorn Hop, p.154

ACTIVITY
DOING

The Disappearing Coin

1. Place a small piece of aluminum foil in a glass filled with water.

2. Position a copper coin, such as a penny, on top of the foil.

3. Let the glass stand for one day and observe what happens.

Describe the appearance of the water at the end of the experiment. Of the aluminum foil. Of the coin.

What chemical reaction has taken place? How do you know?

the analogy of two couples playing doubles in tennis who switch partners.

CONTENT DEVELOPMENT

Ask students if they have ever taken medicine for an upset stomach. Then have them speculate on what chemicals might have been in the medicine.

● ● ● ● **Integration** ● ● ● ●

Use the discussion of magnesium carbonate to integrate digestion into your lesson.

When you take the cap off a bottle of soda, bubbles rise quickly to the top. Why? Carbonated beverages such as soda contain the compound carbonic acid, H_2CO_3. This compound decomposes into water (H_2O) and carbon dioxide gas (CO_2). The CO_2 gas makes up the bubbles that are released. The balanced equation for the decomposition of carbonic acid is

$$H_2CO_3 \longrightarrow H_2O + CO_2$$

Carbonic acid $\longrightarrow$ Water + Carbon dioxide

Single-Replacement Reaction

In a **single-replacement reaction**, an uncombined element replaces an element that is part of a compound. These reactions take the form:

$$A + BX \longrightarrow AX + B$$

Notice that the atom represented by the letter X switches its partner from B to A.

An example of a single-replacement reaction is the reaction between sodium and water. The very active metal sodium must be stored in oil, not water. When it comes in contact with water, it reacts explosively. The sodium replaces the hydrogen in the water and releases lots of energy. The balanced equation for the reaction of sodium with water is:

$$2Na + 2H_2O \longrightarrow 2NaOH + H_2$$

Sodium + Water $\longrightarrow$ Sodium hydroxide + Hydrogen

Most single-replacement reactions, however, do not cause explosions.

INDEPENDENT PRACTICE

▶ *Activity Book*

Students who need practice with the concepts of this section should be provided with the Chapter 2 activity called Types of Chemical Reactions.

ENRICHMENT

Copper replaces silver in silver nitrate because copper is the more active metal. Have students use research materials in the classroom or library to find out what is meant by the activity series of metals.

Double-Replacement Reaction

In a **double-replacement reaction**, different atoms in two different compounds replace each other. In other words, two compounds react to form two new compounds. These reactions take the form:

$$AX + BY \longrightarrow AY + BX$$

Notice that in this reaction the atoms represented by both the letters X and Y switch partners.

If you have ever had an upset stomach, you may have taken a medicine that contained the compound magnesium carbonate. This compound reacts with the hydrochloric acid in your stomach in the following way:

$$MgCO_3 + 2\ HCl \longrightarrow MgCl_2 + H_2CO_3$$

Magnesium carbonate
+ Hydrochloric acid $\longrightarrow$ Magnesium chloride
+ Carbonic acid

In this double-replacement reaction, the magnesium and hydrogen replace each other, or switch partners. One product is magnesium chloride, a harmless compound. The other product is carbonic acid. Do you remember what happens to carbonic acid? It decomposes into water and carbon dioxide. Your stomachache goes away because instead of too much acid, there is now water and carbon dioxide. You owe your relief to this double-replacement reaction:

$$MgCO_3 + 2HCl \longrightarrow MgCl_2 + H_2O + CO_2$$

Magnesium carbonate
+ Hydrochloric acid $\longrightarrow$ Magnesium chloride +
Water + Carbon dioxide

Figure 2–10 *Because copper is a more active metal than silver, it can replace the silver in silver nitrate. In these four photos, you can see the gradual buildup of silver metal on the coil. What type of reaction is this? What other indication is there that a chemical change is taking place?* ❶

ACTIVITY
DISCOVERING

Double-Replacement Reaction

1. Place a small amount of baking soda in a glass beaker or jar.

2. Pour some vinegar on the baking soda. Observe what happens.

Baking soda is sodium hydrogen carbonate, $NaHCO_3$. Vinegar is acetic acid, $HC_2H_3O_2$. Write the chemical equation for this reaction. What gas is produced?

■ How could you test for the presence of this gas?

O ■ 49

2–4 Energy of Chemical Reactions

Figure 2–11 *Paints are chemical compounds produced by double-replacement reactions. What is the general form of a double-replacement reaction?* ①

Figure 2–12 *The explosion of a firecracker is an exothermic reaction. Why is this one reason that firecrackers are dangerous?* ②

50 ■ O

2–3 Section Review

1. Name the four types of reactions.
2. What is the difference between a synthesis reaction and a decomposition reaction?
3. What is a single-replacement reaction? A double-replacement reaction?

Critical Thinking—*Identifying Reactions*

4. What type of reaction is represented by each of the following equations:
   a. $CaCO_3 \longrightarrow CaO + CO_2$
   b. $C + O_2 \longrightarrow CO_2$
   c. $BaBr_2 + K_2SO_4 \longrightarrow 2KBr + BaSO_4$

2–4 Energy of Chemical Reactions

Energy is always involved in a chemical reaction. Sometimes energy is released, or given off, as the reaction takes place. Sometimes energy is absorbed. **Based on the type of energy change involved, chemical reactions are classified as either exothermic or endothermic reactions.**

In either type of reaction, energy is neither created nor destroyed. It merely changes position or form. The energy released or absorbed usually takes the form of heat or visible light.

Exothermic Reactions

A chemical reaction in which energy is released is an **exothermic** (ek-soh-THER-mihk) **reaction.** The word exothermic comes from the root *-thermic,* which refers to heat, and the prefix *exo-,* which means out of. Heat comes out of, or is released from, a reacting substance during an exothermic reaction. A reaction that involves burning, or a combustion reaction, is an example of an exothermic reaction. The combustion of methane gas, which occurs in a gas stove, releases a large amount of heat energy.

The energy that is released in an exothermic reaction was originally stored in the molecules of the

---

eactants. Because the energy is released during the eaction, the molecules of the products do not re- eive this energy. So the energy of the products is ess than the energy of the reactants. Energy diagrams, such as the ones shown in Figure 2–14, an be used to show the energy change in a reac- ion. Note that in an exothermic reaction, the reac- ants are higher in energy than the products are.

Endothermic Reactions

A chemical reaction in which energy is absorbed s an **endothermic reaction**. The prefix *endo-* means nto. During an endothermic reaction, energy is tak- n into a reacting substance. The energy absorbed luring an endothermic reaction is usually in the orm of heat or light. The decomposition of sodium hloride, or table salt, is an endothermic reaction. t requires the absorption of electric energy.

The energy that is absorbed in an endothermic eaction is now stored in the molecules of the prod- ucts. So the energy of the products is more than the nergy of the reactants. See Figure 2–14 again.

Activation Energy

The total energy released or absorbed by a chem- cal reaction does not tell the whole story about the nergy changes involved in the reaction. In order or the reactants to form products, the molecules of he reactants must combine to form a short-lived, igh-energy, extremely unstable molecule. The atoms f this molecule are then rearranged to form prod- ucts. This process requires energy. The molecules of

Figure 2-13 *The cooking of pancakes is an endothermic reaction. Why?*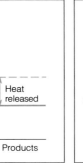

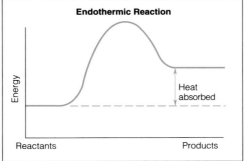

Ⓐctivity Bank

Toasting to Good Health, p.155

Figure 2–14 *An energy diagram for an exothermic reaction indicates that heat is released during the reaction. Heat is absorbed during an endothermic reaction, as shown by its energy diagram. How does the heat content of products and reactants compare for each type of reaction?*

Exothermic Reaction

Heat released

eactants   Products

Endothermic Reaction

Energy

Heat absorbed

Reactants   Products

energy released by the formation of bonds, the reaction is endothermic.

● ● ● ● **Integration** ● ● ● ●

Use the derivation of the word *exother- mic* to integrate language arts into your lesson.

REINFORCEMENT/RETEACHING

Some students may have trouble dis- tinguishing between the words *endothermic* and *exothermic*. Suggest that they try this memory trick: *Endo* sounds like *into,* so it means energy goes *into* the reaction. *Exo* resembles the word *exit,* so it means en- ergy goes *out of* the reaction.

CONTENT DEVELOPMENT

• **How can you tell that an energy change is occurring during a chemical reaction?** (You may see light or feel heat. In some cases, an electric current is produced. Or you may realize that heat, light, or elec- tricity is being absorbed by substances un- dergoing a chemical change.)

Emphasize that a reaction that releases energy is called an exothermic reaction and that a reaction that absorbs energy is called an endothermic reaction

● ● ● ● **Integration** ● ● ● ●

Use the discussion of energy release in endothermic reactions to integrate con- cepts of energy changes into your lesson.

TEACHING STRATEGY 2–4

FOCUS/MOTIVATION

Write the following phrases on the chalkboard:

Striking a match
Setting off a flashbulb
Setting off a firecracker
• **The chemical reactions produced by each of these actions all have something in common. What is it?** (Answers may vary,

but guide students to realize that all the reactions produce heat and light.)

CONTENT DEVELOPMENT

Point out that chemical reactions in- volve the breaking and forming of chem- ical bonds. The breaking of bonds re- quire energy, and the formation of bonds releases energy. When the energy required to break bonds is less than the energy re- leased by the formation of bonds, the re- action is exothermic. When the energy re- quired to break bonds is greater than the

ACTIVITY
THINKING
KITCHEN CHEMISTRY

Skills: Making observations, applying concepts, making inferences

This activity will help students relate the somewhat esoteric topics in their textbook to their everyday world. Check students' charts carefully for accuracy.

2–4 (continued)

CONTENT DEVELOPMENT

Explain that activation energy is the energy level that must be reached in order for a reaction to occur. Emphasize that for both endothermic and exothermic reactions, the activation energy is always higher than the initial energy of the reactants.

GUIDED PRACTICE

Skills Development
Skill: Interpreting charts

Have students study the energy diagrams in Figure 2–15. Explain that the horizontal axis of the graph shows the passage of time as the reaction progresses.

Display the diagrams using an overhead projector. Trace with a pointer the energy path of the reactants as they change into products. Stress that energy is stored in the molecules of products and reactants.

• **In the exothermic reactions, which has more stored energy: the molecules of the reactants or the molecules of the products? How can you tell?** (The molecules of the reactants. The energy curve at the

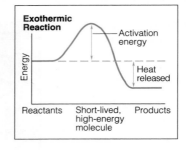
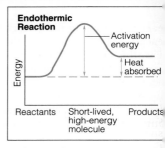

Figure 2–15 *As you can see by these energy diagrams, both an exothermic reaction and an endothermic reaction require activation energy.*

Exothermic Reaction — Activation energy — Heat released

Energy

Reactants — Short-lived, high-energy molecule — Products

Endothermic Reaction — Activation energy — Heat absorbed

Energy

Reactants — Short-lived, high-energy molecule — Products

the reactants must "climb" to the top of an "energy hill" before they can form products. The energy needed to "climb" to the top of the "energy hill" is called **activation energy**. After the reactants have absorbed this activation energy, they can "slide down" the energy hill to form products.

An energy diagram indicates more than whether a reaction is exothermic or endothermic. An energy diagram shows the activation energy of the reaction. Figure 2–15 shows an energy diagram for both an exothermic reaction and an endothermic reaction.

All chemical reactions require activation energy. Even an exothermic reaction such as the burning of a match requires activation energy. In order to light a match, it must first be struck. The friction of the match against the striking pad provides the necessary activation energy.

ACTIVITY
THINKING

Kitchen Chemistry

Many interesting chemical reactions occur during various cooking processes. Observe someone preparing and cooking different kinds of food. Record your observations of the changes that take place in the properties of the food. Are the changes physical or chemical? Endothermic or exothermic? Synthesis, decomposition, or replacement?

Using books and other reference materials in the library, find out more about each of the chemical reactions you have listed.

2–4 Section Review

1. What is an exothermic reaction? An endothermic reaction?
2. On which side should the energy term be written in an equation representing an endothermic reaction? In an equation representing an exothermic reaction?
3. Compare the energy content of reactants and products in an exothermic reaction. In an endothermic reaction.

Connection—*You and Your World*
4. Using what you know about activation energy, explain why a match will not light if it is not struck hard enough.

---

end of the reaction is lower than the curve at the beginning of the reaction.)

• **Which molecules have the greater amount of stored energy in the endothermic reaction? How can you tell?** (The products. The curve at the end of the reaction is higher than the curve at the beginning of the reaction.)

INDEPENDENT PRACTICE
Section Review 2–4

1. An exothermic reaction is a chemical reaction in which energy is released. An endothermic reaction is a chemical reaction in which energy is absorbed.

2. On the left with the reactants. On the right with the products.

3. Reactants have more energy than products in an exothermic reaction. Products have more energy than reactants in an endothermic reaction.

4. If a match is not struck hard enough, it will not light because the friction energy created in the strike will not be enough to start the reaction.

2–5 Rates of Chemical Reactions

The complete burning of a thick log can take many hours. Yet if the log is ground into fine sawdust, the burning can take place at dangerously high speeds. In fact, if the dust is spread through the air, the burning can produce an explosion! In both these processes, the same reaction is taking place. The various substances in the wood are combining with oxygen. One reaction, however, proceeds at a faster speed than the other one does. What causes the differences in reaction times?

In order to explain differences in reaction time, chemists must study **kinetics**. Kinetics is the study of **reaction rates**. The rate of a reaction is a measure of how quickly reactants turn into products. Reaction rates depend on a number of factors, which you will now read about.

Collision Theory

You learned that chemical reactions occur when bonds between atoms are broken, the atoms are re-arranged, and new bonds are formed. In order for this process to occur, particles must collide. As two particles approach each other, they begin to interact. During this interaction, old bonds may be broken and new bonds formed. For a reaction to occur, however, particles must collide at precisely the correct angle with the proper amount of energy. The more collisions that occur under these conditions, the faster the rate of the chemical reaction.

A theory known as the **collision theory** relates particle collisions to reaction rate. **According to the collision theory, the rate of a reaction is affected by four factors: concentration, surface area, temperature, and catalysts.**

Concentration

The concentration of a substance is a measure of the amount of that substance in a given unit of volume. A high concentration of reactants means there

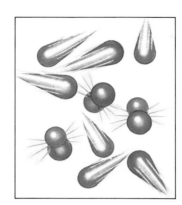

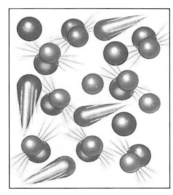

Figure 2–16 *Collisions of molecules increase when there are more molecules. The number of molecules per unit volume is called concentration. What is the relationship between the concentration of reactants and reaction rate?* ❶

O ■ 53

2–5 Rates of Chemical Reactions

MULTICULTURAL OPPORTUNITY 2–5

To help students understand the collision theory, conduct the following activity. Rope off an area of your classroom about 2 meters square. Place two students in that roped area and have them close their eyes and walk around. Every time they touch the rope, they are to turn around. Count the number of times the students collide. Gradually increase the number of students until the area is full. Continue to monitor the number of collisions. Make the analogy between this activity and the relationship between concentration and reaction time.

ESL STRATEGY 2–5

Have ESL students work with English-speaking partners to prepare answers to the following questions:
1. What must happen for a chemical reaction to occur? (The bonds between atoms are broken, the atoms are rearranged, and new bonds are formed.)
2. What is the name of the study of reaction rates? (Kinetics.)
3. What are four things that affect the rate of chemical reactions? (The concentration of a substance, the surface area, the temperature, and the addition of a catalyst.)
4. What substance increases the rate of a chemical reaction but is not itself changed by the reaction? (A catalyst.)

REINFORCEMENT/RETEACHING

Review students' responses to the Section Review questions. Reteach any material that is still unclear, based on students' responses.

CLOSURE

▶ *Review and Reinforcement Guide*
Have students complete Section 2–4 in the *Review and Reinforcement Guide.*

TEACHING STRATEGY 2–5

FOCUS/MOTIVATION

Have students consider the following situations and relate them to the collision theory:
• **Suppose you are walking around a department store on a day when the store is nearly empty. What are the chances of your bumping into another person?** (There is almost no chance.)
• **Suppose you are walking around the**

same store on a day when the store is moderately busy. What are the chances of your bumping into another person? (There is a reasonable chance, maybe about 50–50.)
• **Now suppose the store is having a giant preholiday sale on a Saturday afternoon. You are trying to get close to the best bargain counter. What are the chances of your bumping into another person?** (There is a very good chance, almost 100 percent.)

Discovery Learning

Skills: Making observations, making comparisons, making inferences

Materials: 2 sugar cubes, 2 plastic cups

In this activity students will note that the powdered sugar dissolves faster than the sugar cube. They should be able to relate that observation to the fact that increasing surface area increases the rate of reaction by increasing the number of collisions between reacting molecules.

2–5 (continued)

CONTENT DEVELOPMENT

Review the idea that molecules are constantly in motion.

• **How is the motion of molecules related to chemical reactions?** (Molecules must collide in order to react; they could not collide if they were not moving.)

• **What condition must be met if colliding molecules are to react?** (They must collide with sufficient energy.)

• **What do you think would have to happen to the collisions in order for a chemical reaction to speed up?** (Collisions would have to happen more frequently or with greater energy.)

Explain that the four factors that affect reaction rate all affect molecular collisions, either by increasing the likelihood of collision or by increasing the energy of molecules.

FOCUS/MOTIVATION

Present students with the following analogies:

• **Suppose you are on a crowded city street. In which situation would you be more likely to bump into someone—when you are strolling in an easy, relaxed way or when you are running to catch a bus?** (When you are running to catch a bus.)

• **Now suppose you are walking at a moderate speed down the same street and someone bumps into you. Would you**

Figure 2–17 *As bellows are pumped, more oxygen is supplied to the fire and the rate of reaction increases. Why are forest fires particularly dangerous when weather conditions nearby include high winds?* ③

ACTIVITY
DISCOVERING

Rate of Reaction

1. Obtain two sugar cubes. Grind one of the cubes into powder.

2. Fill two clear-plastic cups with warm tap water.

3. As close to the same time as possible, put the whole sugar cube into one cup of water and the powdered sugar into the other.

4. Stir each cup briefly every 30 seconds. Observe the time required for the complete dissolving of sugar in each cup.

How do the times compare? Which reaction rate is faster?

■ What important relationship regarding reaction rate have you discovered?

54 ■ O

are a great many particles per unit volume. So there are more particles of reactants available for collisions. More collisions occur and more products are formed in a certain amount of time. What does a low concentration of reactants mean? ①

Generally, most chemical reactions proceed at a faster rate if the concentration of the reactants is increased. A decrease in the concentration of the reactants decreases the rate of reaction. For example, a highly concentrated solution of sodium hydroxide (NaOH), or lye, will react more quickly to clear a clogged drain than will a less concentrated lye solution. Why would the rate of burning charcoal be increased by blowing on the fire? ②

Surface Area

When one of the reactants in a chemical reaction is a solid, the rate of reaction can be increased by breaking the solid into smaller pieces. This increases the surface area of the reactant. Surface area refers to how much of a material is exposed. An increase in surface area increases the collisions between reacting particles.

A given quantity of wood burns faster as sawdust than it does as logs. Sawdust has a much greater surface area exposed to air than do the logs. So oxygen particles from the air can collide with more wood particles per second. The reaction rate is increased.

Figure 2–18 *It may seem hard to believe that grains of wheat or even dust can cause the immense destruction you see here. But an explosion at a grain elevator is an ever-present danger because a chemical reaction can occur almost instantaneously. What reaction-rate factor is responsible for such an explosion?* ④

rather that person be walking slowly or running full speed? Why? (Walking slowly. The collision will have much less impact at a slower speed.)

Use these analogies to point out to students that when the temperature of molecules increases, the molecules begin to move faster, just as the person running is moving faster than the person walking. Thus, the likelihood of collision increases, as does the force of collision.

GUIDED PRACTICE

Skills Development

Skills: Applying concepts, making observations, making comparisons

At this point have students complete the in-text Chapter 2 Laboratory Investigation: Determining Reaction Rate. In the investigation students will determine how concentration affects reaction rate.

Many medicines are produced in the form of a fine powder or many small crystals. Medicine in this form is often more effective than the same medicine in tablet form. Do you know why? Tablets dissolve in the stomach and enter the bloodstream at a slower rate. How does the collision theory account for the fact that fine crystals of table salt dissolve more quickly in water than do large crystals of rock salt? ⑤

Temperature

An increase in temperature generally increases the rate of a reaction. Here again the collision theory provides an explanation for this fact. Particles are constantly in motion. Temperature is a measure of the energy of their motion. Particles at a high temperature have more energy of motion than do particles at a low temperature. Particles at a high temperature move faster than do particles at a low temperature. So particles at a high temperature collide more frequently. They also collide with greater energy. This increase in the rate and energy of collisions affects the reaction rate. More particles of reactants are able to gain the activation energy needed to form products. So reaction rate is increased.

At room temperature, the rates of many chemical reactions roughly double or triple with a rise in temperature of 10°C. How does this fact explain the use of refrigeration to keep foods from spoiling? ⑥

Catalysts

Some chemical reactions take place very slowly. The reactions involved with digesting a cookie are examples. In fact, if these reactions proceeded at their normal rate, it could take weeks to digest one cookie! Fortunately, certain substances speed up the rate of a chemical reaction. These substances are catalysts. A catalyst is a substance that increases the rate of a reaction but is not itself changed by the reaction. Although a catalyst alters the reaction, it can be recovered at the end of the reaction.

How does a catalyst change the rate of a reaction if it is not changed by the reaction? Again, the explanation is based on the collision theory. Reactions often involve a series of steps. A catalyst changes one

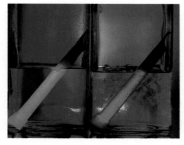

Figure 2–19 *These glow-in-the-dark sticks are called Cyalume light sticks. When a light stick is placed in hot water (left), it glows more brightly than when placed in cold water (right). The reaction that causes a stick to glow is faster and produces more light at higher temperatures.*

ACTIVITY
DISCOVERING

Temperature and Reaction Rate

1. Fill one glass with cold water and another with hot water.

2. Drop a seltzer tablet into each glass of water and observe the reactions that occur.

Is there any noticeable difference in the two reactions?

■ What relationship regarding reaction rate does this activity illustrate?

■ Does this activity prove that a temperature difference always has the same effect?

O ■ 55

ACTIVITY
DISCOVERING
TEMPERATURE AND REACTION RATE

Discovery Learning

Skills: Making observations, making comparisons, making inferences

Materials: 2 glasses, 2 seltzer tablets

In this activity students observe that the reaction rate of the seltzer tablet increases in hot water versus cold water. They will be able to infer that the increase in temperature increased the reaction rate. Caution students, however, that increasing temperature does not always increase reaction rate.

FOCUS/MOTIVATION

Have students consider the following question:

• **In cooking, what factors would tend to affect the rates of the chemical reactions taking place? How might the finished product be ruined if these factors were not properly taken into account?** (Probably the two most important factors in cooking are temperature and surface area. A recipe calls for a certain temperature to ensure that the reactions do not take place too slowly or too quickly. A good example of food ruined by a too-slow cooking temperature is baking bread—the dough will not rise sufficiently. Surface area comes into play when a recipe calls for ingredients to be chopped, ground, or melted. An example of this is adding sugar to a recipe—if sugar cubes were added instead of granulated sugar, the finished product would probably have many lumps of unreacted sugar!)

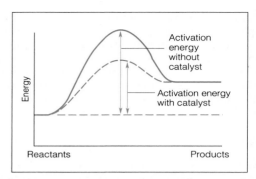

Figure 2–20 *A catalyst changes the rate of a chemical reaction without itself being changed by the reaction. According to this energy diagram, how does a catalyst affect the rate of a reaction?* ①

Figure 2–21 *This collection of colorful pellets actually contains catalysts used in industry. The blue catalyst in front, for example, is* ② *used in the reactions that remove sulfur and nitrogen from crude oil.*

or more of the steps. A catalyst produces a different, lower energy path for the reaction. In other words, it lowers the "energy hill," or activation energy. A decrease in the activation energy needed for the reaction allows more reactant particles to form products. Collisions need not be so energetic. Therefore, more collisions are successful at producing products.

A catalyst is usually involved in one or more of the early steps in a reaction. The catalyst is, however, re-formed during a later step. This explains why a catalyst can be recovered at the end of the reaction.

Catalysts are used in many chemical processes. Certain automobiles contain devices called catalytic converters. A catalytic converter speeds up the reaction that changes the harmful gases produced by automobile engines into harmless ones. Some of the most important catalysts are those found in your body. Catalysts in the body are called enzymes. Each enzyme increases the rate of a specific reaction involved in the body's metabolism.

2–5 Section Review

1. What is reaction rate?
2. How is reaction rate related to collision theory?
3. Name four factors that affect reaction rate.
4. How does collision theory explain the effect of a decrease in temperature on the reaction rate?

Critical Thinking—*Drawing Conclusions*
5. Do you think there is a need for catalysts that slow down a reaction? Give some examples.

CONNECTIONS

❸ *Chemical Reactions That Destroy the Environment*

Stop and look around for a moment. Do you notice that you are surrounded by all sorts of devices that make life easier and more comfortable? There are cars, airplanes, air conditioners, televisions, refrigerators, radios, and industrial machinery—to mention a few. Where do all these inventions get their energy? Most of it comes from the burning of coal and petroleum.

Coal and petroleum are fuels that contain the element carbon. When they burn (a combustion reaction), carbon dioxide (CO_2) is released. The combustion of coal and petroleum releases about 20 billion tons of carbon dioxide into the atmosphere every year. And carbon dioxide in the atmosphere is of great consequence. The Earth and its inhabitants send a great deal of heat out through the atmosphere toward space. It is vital that this heat escapes the Earth's atmosphere. If it does not, the surface of the Earth will heat up, changing climates significantly. Although carbon dioxide is only a small part of the atmosphere, it has the ability to trap heat near the surface of the Earth. By absorbing heat released from the Earth, carbon dioxide acts like the glass over a greenhouse. For this reason, the warming of the Earth by carbon dioxide and other heat-trapping gases is referred to as the

greenhouse effect. Some climatologists (scientists who study climate) believe that the buildup of carbon dioxide and other gases has already begun to change the climate of our planet. They cite as evidence the fact that the six warmest years on record were all in the 1980s. The trend is continuing into the 1990s.

One long-term outcome of the warming of the Earth (global warming) is the melting of the polar icecaps. As the icecaps melt because of higher temperatures, the sea level rises, causing certain coastal areas to be flooded and destroyed. Another effect is a shift in the distribution of rainfall over the various continents. Eventually, heat and drought could become commonplace over the vast region of the United States that now provides most of the world's agricultural products.

In order to minimize and possibly reverse the greenhouse effect, new energy sources that do not release carbon dioxide into the atmosphere must be developed. Energy must also be conserved. What role do you think you can play in making sure chemical reactions do not further destroy the environment?

O ■ 57

CONNECTIONS

CHEMICAL REACTIONS THAT DESTROY THE ENVIRONMENT

After students have read and discussed the Connections feature, you may wish to introduce the topic of alternative sources of energy. These might include hydroelectric power, power from the tides or the wind, geothermal energy, and solar heat.

Hydroelectic power uses the kinetic energy of falling water to turn a turbine and thus makes use of a fuel that is free and not subject to inflation. Tidal power is generated from the flow of water caused by the rotation of the Earth and the gravitational pull of both the sun and the moon. Windmills produce power when the wind pushes on the blades. Geothermal heat is energy from the interior of the Earth in the form of hot water or steam. And solar energy includes a number of solar options that are based on collecting and using energy from the sun.

Although each of these alternative forms of energy production presents its own environmental challenges, it seems likely that one or more of them will be less destructive than current practices are.

If you are teaching thematically, you may want to use the Connections feature to reinforce the themes of energy systems and interactions.

Integration: Use the Connections feature to integrate the atmosphere into your science lesson.

REINFORCEMENT/RETEACHING

▶ *Activity Book*

Students who have difficulty understanding the concepts of this section should be provided with the Chapter 2 activity called Testing Reaction Rate Factors.

INDEPENDENT PRACTICE

Section Review 2–5

1. Reaction rate is a measure of how quickly reactants turn into products.

2. According to the collision theory, molecules cannot react unless they collide with sufficient energy. The faster and stronger the collisions, the faster the reaction rate.

3. Four factors that affect reaction rate are temperature, concentration, surface area, and the presence of a catalyst.

4. Collision theory assumes that when temperature is decreased, molecular motion slows down. This, in turn, decreases the number of collisions and slows down the reaction rate.

5. Accept all logical answers. Students might suggest that catalysts that slow down the rate of food spoilage and rust would be very helpful. Catalysts that slow down the rate of reactions are called negative catalysts, or inhibitors.

REINFORCEMENT/RETEACHING

Review students' responses to the Section Review questions. Reteach any material that is still unclear, based on students' responses.

CLOSURE

▶ *Review and Reinforcement Guide*

Have students complete Section 2–5 in the *Review and Reinforcement Guide*.

Laboratory Investigation

DETERMINING REACTION RATE

BEFORE THE LAB

1. Gather all materials at least one day prior to the investigation. You should have enough supplies to meet your class needs, assuming six students per group.
2. Prepare solutions A and B as follows: For Solution A, dissolve 4.3 g potassium iodate (KIO_3) per liter of water. For every liter of Solution B, make a paste of 4 g soluble starch in a small amount of water. Add paste slowly to 900 mL boiling water. Boil for several minutes and allow to cool. Just before use, add 0.2 g $Na_2S_2O_5$ and 5 mL of 1 molar H_2SO_4.

PRE-LAB DISCUSSION

Have students read the complete laboratory procedure. Discuss the procedure by asking questions similar to the following:
• **Which of these factors is being tested in this lab?** (Concentration.)
• **What is the variable?** (The concentration of Solution A.)
• **Which factor is the control?** (Concentration of Solution B.)
• **Can you offer a hypothesis about the outcome of this experiment?** (Answers may vary; the correct hypothesis is that the rate of reaction will vary directly with the concentration of Solution A.)

SAFETY TIPS

Emphasize that because Solution B contains a strong acid, it must be handled carefully. Caution students to pour slowly to avoid splashing and to wear their safety goggles throughout the investigation.

Laboratory Investigation

Determining Reaction Rate

Problem

How does concentration affect reaction rate?

Materials (per group)

safety goggles
2 graduated cylinders
120 mL Solution A
3 250-mL beakers
distilled water at room temperature
90 mL Solution B
stirring rod
sheet of white paper
stopwatch or watch with a sweep second hand

Procedure 🧪 📷 👁

1. Carefully measure 60 mL of Solution A and pour it into a 250-mL beaker. Add 10 mL of distilled water and stir.
2. Carefully measure 30 mL of Solution B and pour it into a second beaker. Place the beaker of Solution B on a sheet of white paper in order to see the color change more easily.
3. Add the 70 mL of Solution A-water mixture to Solution B. Stir rapidly. Record the time it takes for the reaction to occur.
4. Rinse and dry the reaction beaker.
5. Repeat the procedure using the other amounts shown in the data table.

Observations

Solution A (mL)	Distilled Water Added to Solution A (mL)	Solution B (mL)	Reaction Time (sec)
60	10	30	
40	30	30	
20	50	30	

1. What visible indication is there that a chemical reaction is occurring?
2. What is the effect of adding more distilled water on the concentration of Solution A?
3. What happens to reaction time as more distilled water is added to Solution A?
4. Make a graph of your observations by plotting time along the X axis and volume of Solution A along the Y axis.

Analysis and Conclusions

1. How does concentration affect reaction rate?
2. Does your graph support your answer to question 1? Explain why.
3. What would a graph look like if time were plotted along the X axis and volume of distilled water added to Solution A were plotted along the Y axis?
4. In this investigation, what is the variable? The control?
5. **On Your Own** Enhance your graph by testing other concentrations.

TEACHING STRATEGIES

1. Have teams follow the directions carefully as they work in the laboratory.
2. Circulate through the classroom, assisting those students who are having difficulty with accurately measuring the solutions.

DISCOVERY STRATEGIES

Discuss how the investigation relates to the chapter ideas by asking open questions similar to the following:

• **What four factors can affect the speed of a chemical reaction?** (The concentration of a substance, the surface area, the temperature, and the presence or absence of a catalyst—relating cause and effect.)
• **Which of these factors are you changing in this experiment?** (The concentration of a substance, Solution A—applying concepts.)
• **How are you changing the concentration of Solution A?** (As students add more water, Solution A becomes less and less concentrated—applying concepts.)

Study Guide

Summarizing Key Concepts

2–1 Nature of Chemical Reactions

▲ When a chemical reaction occurs, there is always a change in the properties and the energy of the substances.

▲ A reactant is a substance that enters into a chemical reaction. A product is a substance that is produced by a chemical reaction.

2–2 Chemical Equations

▲ The law of conservation of mass states that matter can be neither created nor destroyed in a chemical reaction.

▲ A chemical equation that has the same number of atoms of each element on both sides of the arrow is a balanced equation.

2–3 Types of Chemical Reactions

▲ In a synthesis reaction, two or more simple substances combine to form a new, more complex substance.

▲ In a decomposition reaction, a complex substance breaks down into two or more simpler substances.

▲ In a single-replacement reaction, an uncombined element replaces an element that is part of a compound.

▲ In a double-replacement reaction, different atoms in two different compounds replace each other.

2–4 Energy of Chemical Reactions

▲ Energy is released in an exothermic reaction. Energy is absorbed in an endothermic reaction.

▲ In order for reactants to form products, activation energy is needed.

2–5 Rates of Chemical Reactions

▲ The rate of a reaction is a measure of how quickly reactants turn into products.

▲ An increase in the concentration of reactants increases the rate of a reaction.

▲ An increase in the surface area of reactants increases the rate of reaction.

▲ An increase in temperature generally increases the rate of a reaction.

Reviewing Key Terms

Define each term in a complete sentence.

2–1 Nature of Chemical Reactions
chemical reaction
reactant
product

2–2 Chemical Equations
chemical equation

2–3 Types of Chemical Reactions
synthesis reaction
decomposition reaction

single-replacement reaction
double-replacement reaction

2–4 Energy of Chemical Reactions
exothermic reaction
endothermic reaction
activation energy

2–5 Rates of Chemical Reactions
kinetics
reaction rate
collision theory
catalyst

O ■ 59

ANALYSIS AND CONCLUSIONS

1. Decreased concentration means decreased reaction rate, which would show up as increased reaction time.
2. Students' graphs should support their answer.
3. The graph would look the same.
4. The variable is the concentration of Solution A. The control is the concentration of Solution B.
5. Students' graphs will vary, depending on concentrations tested. But in general, the graphs should have the same basic shape.

GOING FURTHER: ENRICHMENT

Part 1
Students may find it interesting to vary the concentration of Solution B rather than the concentration of Solution A. Have students offer hypotheses before the investigation; then discuss the outcome.

Part 2
Have students design an experiment to test the effect of temperature on the reaction between solutions A and B. Note that it is often safer to test the effects of temperature by cooling rather than by heating because some reactants can be hazardous to handle at high temperatures.

• **What effect would you expect this to have on the speed of the chemical reaction occurring between Solution A and Solution B?** (It should slow it down—relating cause and effect.)

• **What else could you do to slow down the chemical reaction?** (Decrease the temperature or add some type of catalyst—making inferences.)

OBSERVATIONS

1. The color change is a visible indication that a chemical reaction is taking place.
2. By adding more distilled water, the concentration is lessened.
3. As most distilled water is added, the reaction time increases because the reaction rate is decreasing.
4. Check students' graphs. They should show a straight line going from the bottom-left corner toward the top-right corner.

Chapter Review

ALTERNATIVE ASSESSMENT

The *Prentice Hall Science* program includes a variety of testing components and methodologies. Aside from the Chapter Review questions, you may opt to use the Chapter Test or the Computer Test Bank Test in your *Test Book* for assessment of important facts and concepts. In addition, Performance-Based Tests are included in your *Test Book*. These Performance-Based Tests are designed to test science process skills, rather than factual content recall. Since they are not content dependent, Performance-Based Tests can be distributed after students complete a chapter or after they complete the entire textbook.

CONTENT REVIEW

Multiple Choice

1. a
2. d
3. a
4. c
5. d
6. d
7. b
8. c

True or False

1. F, products
2. T
3. F, decomposition
4. F, synthesis
5. F, less
6. T
7. T

Concept Mapping

Row 1: Products, Coefficients
Row 2: The law of conservation of mass

CONCEPT MASTERY

1. A chemical reaction is a process in which the chemical properties of the original substances disappear as new substances with different chemical properties are formed. Substances react chemically because the atoms of the substances have complementary bonding capacities. The atoms of a substance form chemical bonds with other atoms in order to complete the outermost energy layers.

Content Review

Multiple Choice

Choose the letter of the answer that best completes each statement.

1. In a balanced chemical equation,
   a. atoms are conserved.
   b. molecules are equal.
   c. coefficients are equal.
   d. energy is not conserved.
2. Two or more substances combine to form one substance in a
   a. decomposition reaction.
   b. double-replacement reaction.
   c. single-replacement reaction.
   d. synthesis reaction.
3. In an endothermic reaction, heat is
   a. absorbed.   c. destroyed.
   b. released.   d. conserved.
4. The energy required for reactants to form products is called
   a. energy of motion.
   b. potential energy.
   c. activation energy.
   d. synthetic energy.
5. The substances to the left of the arrow in a chemical equation are called
   a. coefficients.   c. subscripts.
   b. products.   d. reactants.
6. An atom's ability to undergo chemical reactions is determined by
   a. protons.   c. innermost electrons.
   b. neutrons.   d. outermost electrons.
7. The rate of a chemical reaction can be increased by
   a. decreasing concentration.
   b. increasing surface area.
   c. removing a catalyst.
   d. all of these.
8. Adding a catalyst to a reaction increases its rate by
   a. increasing molecular motion.
   b. decreasing molecular motion.
   c. lowering activation energy.
   d. increasing concentration.

True or False

If the statement is true, write "true." If it is false, change the underlined word or words to make the statement true.

1. The substances formed as a result of a chemical reaction are called <u>reactants</u>.
2. A number written in front of a chemical symbol or formula is a(an) <u>coefficient</u>.
3. In a <u>synthesis</u> reaction, complex substances form simpler substances.
4. The formation of carbon dioxide during combustion of a fuel is an example of a <u>decomposition</u> reaction.
5. In an exothermic reaction, products have <u>more</u> energy than reactants.
6. The study of the rates of chemical reactions is <u>kinetics</u>.
7. The <u>collision theory</u> can be used to account for the factors that affect reaction rates.

Concept Mapping

Complete the following concept map for Section 2–2. Refer to pages O6–O7 to construct a concept map for the entire chapter.

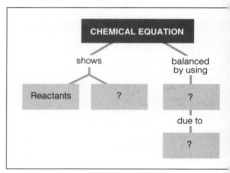

2. A chemical equation is an expression in which symbols and formulas are used to represent a chemical reaction.

3. Mass remains constant in a chemical reaction; that is, the mass of the products equals the mass of the reactants.

4. A coefficient is a number placed in front of a chemical symbol in order to balance a chemical equation. H_2O_2 represents 2 hydrogen atoms and 2 oxygen atoms. In contrast, $2H_2O$ shows two molecules, each containing 2 hydrogen atoms and 1 oxygen atom.

5. Synthesis: Substances combine to form new substances, $2Na + Cl_2 \rightarrow 2NaCl$. Decomposition: Complex substance breaks down into two or more substances, $H_2CO_3 \rightarrow H_2O + CO_2$. Single-replacement: Uncombined element replaces an element that is part of a compound, $2Na + 2H_2O \rightarrow 2NaOH + H_2$. Double-replacement: Two compounds react to form two new compounds, $MgCO_3 + 2HCl \rightarrow MgCl_2 + H_2CO_3$.

6. Activation energy is energy that must be added to cause a chemical reaction to oc-

Concept Mastery

Discuss each of the following in a brief paragraph.

1. What is a chemical reaction? Why do substances react chemically?
2. Describe what a chemical equation is.
3. State the law of conservation of mass and explain its role in chemical equations.
4. What is a coefficient? What is the difference between H_2O_2 and $2H_2O$?
5. Describe each of the four types of chemical reactions. Include an equation showing the general form for each.
6. What is activation energy? How is it affected by a catalyst?
7. Explain how the energy content of the products of a reaction compares with that of the reactants when the reaction is exothermic. When it is endothermic.
8. Use the collision theory to explain the effects on reaction rate of increased concentration, catalysts, and increased surface area.
9. Give two reasons why collisions between molecules of reactants may not be effective in forming products.

Critical Thinking and Problem Solving

Use the skills you have developed in this chapter to answer each of the following.

1. **Making calculations** Balance the following equations:
   a. $PbO_2 \longrightarrow PbO + O_2$
   b. $Ca + H_2O \longrightarrow Ca(OH)_2 + H_2$
   c. $Zn + S \longrightarrow ZnS$
   d. $BaCl_2 + Na_2SO_4 \longrightarrow BaSO_4 + NaCl$
   e. $Al + Fe_2O_3 \longrightarrow Al_2O_3 + Fe$
   f. $C_{12}H_{22}O_{11} \longrightarrow C + H_2O$
2. **Classifying reactions** Identify the general type of reaction represented by each equation. Explain your answers.
   a. $Fe + 2HCl \longrightarrow FeCl_2 + H_2$
   b. $NiCl_2 \longrightarrow Ni + Cl_2$
   c. $4C + 6H_2 + O_2 \longrightarrow 2C_2H_6O$
   d. $2LiBr + Pb(NO_3)_2 \longrightarrow 2LiNO_3 + PbBr_2$
   e. $CaO + H_2O \longrightarrow Ca(OH)_2$
3. **Developing a model** Draw an energy diagram of an exothermic reaction that has a high activation energy. On your diagram, indicate how an increase in temperature would affect the rate of this reaction. Do the same for the addition of a catalyst.
4. **Applying concepts** Why can a lump of sugar be used in hot tea but granulated (small crystals) sugar is preferred in iced tea?

5. **Recognizing relationships** Potential energy is energy that is stored for later use. Explain where the potential energy is located in an exothermic reaction? Why do foods and fuel have chemical potential energy?
6. **Using the writing process** You are hired by a large company to design an advertising campaign. The company galvanizes, or covers, iron with the more active metal zinc to protect iron from corroding. Develop an advertising slogan and poster that explain to the public why this process is important.

---

cur. A catalyst decreases the amount of activation energy required for a chemical reaction to take place.

7. In an exothermic reaction, products have less energy than reactants. In an endothermic reaction, products have more energy than reactants.

8. The reaction rate is increased.

9. The collisions may not be energetic enough or numerous enough.

CRITICAL THINKING AND PROBLEM SOLVING

1a. $2PbO_2 \rightarrow 2PbO + O_2$
b. $Ca + 2H_2O \rightarrow Ca(OH)_2 + H_2$
c. $Zn + S \rightarrow ZnS$
d. $BaCl_2 + Na_2SO_4 \rightarrow BaSO_4 + 2NaCl$
e. $2Al + Fe_2O_3 \rightarrow Al_2O_3 + 2Fe$
f. $C_{12}H_{22}O_{11} \rightarrow 12C + 11H_2O$
2a. single replacement, **b.** decomposition, **c.** synthesis, **d.** double replacement, **e.** synthesis.

---

3. Check students' diagrams. They should note that an increase in temperature as well as a catalyst will lower the height of the activation "energy hill."

4. The temperature of the hot tea helps to dissolve the lump of sugar. For the iced tea, the smaller sugar crystals have a greater surface area and thus are dissolved more quickly than a lump of sugar would be.

5. The energy that is released in an exothermic reaction was originally stored in the molecules of the reactants. The stored energy in foods and fuel can be released in exothermic reactions.

6. Students' advertising campaigns should be scientifically accurate. They may want to include a simple explanation of the rusting of iron. The iron reacts with oxygen in the air, slowly releasing heat and producing iron oxide, or rust. Removing the rust can be expensive, involving dipping the iron parts in a sulfuric acid bath. It is better to prevent the rust than to have to remove it later.

KEEPING A PORTFOLIO

You might want to assign some of the Concept Mastery and Critical Thinking and Problem Solving questions as homework and have students include their responses to unassigned questions in their portfolio. Students should be encouraged to include both the question and the answer in their portfolio.

ISSUES IN SCIENCE

The following issue can be used as a springboard for discussion or given as a writing assignment:

A tremendous amount of money was spent on the restoration of the Statue of Liberty. Much of this money was raised by a foundation comprising private citizens who devoted a great amount of time and energy to the cause. Do you think such a great expenditure of money and effort was worthwhile, or could it have been better spent on other things? Take a stand on this issue and present arguments to support your position.

Chapter 3 FAMILIES OF CHEMICAL COMPOUNDS

SECTION	HANDS-ON ACTIVITIES
3–1 Solution Chemistry pages O64–O72 Multicultural Opportunity 3–1, p. O64 ESL Strategy 3–1, p. O64	**Student Edition** ACTIVITY (Doing): Emulsifying Action, p. O66 ACTIVITY (Discovering): Solubility of a Gas in a Liquid, p. O69 ACTIVITY (Discovering): Sudsing Power, p. O70 **Laboratory Manual** Relating Solubility and Temperature, p. O35 **Activity Book** CHAPTER DISCOVERY: What's in a Flavor? p. O59 Product Testing Activity Testing Bottled Water **Teacher Edition** Observing the Conductivity of Solutions, p. O62d
3–2 Acids and Bases pages O72–O75 Multicultural Opportunity 3–2, p. O72 ESL Strategy 3–2, p. O72	**Student Edition** ACTIVITY (Doing): Acid-Base Testing, p. O73 ACTIVITY BANK: In a Jam, p. O156 **Laboratory Manual** Testing Unknown Substances Using Acid-Base Indicators, p. O39 Testing Commercial Antacids, p. O43 Acids and Bases From Home, p. O45
3–3 Acids and Bases in Solution: Salts pages O76–O78 Multicultural Opportunity 3–3, p. O76 ESL Strategy 3–3, p. O76	**Student Edition** ACTIVITY (Doing): A Homemade Indicator, p. O77 LABORATORY INVESTIGATION: Acids, Bases, and Salts, p. O92 **Activity Book** ACTIVITY BANK: Soda Fountain, p. O157 **Product Testing Activity** Testing Antacids
3–4 Carbon and Its Compounds pages O78–O82 Multicultural Opportunity 3–4, p. O78 ESL Strategy 3–4, p. O78	**Student Edition** ACTIVITY (Doing): Octane Rating, p. O79 ACTIVITY (Doing): Isomers, p. O81
3–5 Hydrocarbons pages O83–O87 Multicultural Opportunity 3–5, p. O83 ESL Strategy 3–5, p. O83	**Laboratory Manual** How Much Vitamin C Is in Fruit Juice? p. O49 **Activity Book** ACTIVITY: Digestion, p. O69 **Product Testing Activity** Testing Cereals
3–6 Substituted Hydrocarbons pages O87–O91 Multicultural Opportunity 3–6, p. O87 ESL Strategy 3–6, p. O87	**Student Edition** ACTIVITY (Discovering): Saturated and Unsaturated Fats, p. O90
Chapter Review pages O92–O95	

OTHER ACTIVITIES	MEDIA AND TECHNOLOGY
Activity Book ACTIVITY: Dissolving Household Substances, p. O65 ACTIVITY: Science Concentration, p. O71 ACTIVITY: Solutions Everywhere, p. O75 ACTIVITY: Solutions and Temperature, p. O77 **Review and Reinforcement Guide** Section 3–1, p. O25	**Video/Videodisc** Elements, Compounds, and Mixtures **English/Spanish Audiotapes** Section 3–1
Activity Book ACTIVITY: Acid, Base, or Neutral Substance? p. O63 **Review and Reinforcement Guide** Section 3–2, p. O29	**Transparency Binder** Acids, Bases, and pH Values **English/Spanish Audiotapes** Section 3–2
Activity Book ACTIVITY: Names and Uses of Salts, p. O61 ACTIVITY: Acid, Base, or Salt? p. O67 **Review and Reinforcement Guide** Section 3–3, p. O31	**Video/Videodisc** Acids, Bases, and Salts **English/Spanish Audiotapes** Section 3–3
Review and Reinforcement Guide Section 3–4, p. O33	**English/Spanish Audiotapes** Section 3–4
Activity Book ACTIVITY: Naming Hydrocarbons, p. O79 **Review and Reinforcement Guide** Section 3–5, p. O35	**English/Spanish Audiotapes** Section 3–5
Review and Reinforcement Guide Section 3–6, p. O37	**English/Spanish Audiotapes** Section 3–6
Test Book Chapter Test, p. O55 Performance-Based Tests, p. O117	**Test Book** Computer Test Bank Test, p. O61

*All materials in the Chapter Planning Guide Grid are available as part of the Prentice Hall Science Learning System.

Chapter 3 FAMILIES OF CHEMICAL COMPOUNDS

CHAPTER OVERVIEW

Solutions are homogeneous mixtures in which molecules are uniformly distributed. Although solutions are sometimes thought of as liquid, there are nine different types of solutions, based on the three phases of matter. Solutions can also be unsaturated, saturated, or supersaturated. Acids, bases, and salts make up one of the most abundant and important families of chemical compounds. The akalinity or acidity of a solution is measured on the pH scale.

Carbon compounds can be found in many things—from ice cream to car wax. The bonding properties of carbon are shown as the basis for organic compounds, and a myriad of substances can be created using carbon because of its bonding ability; for example, with hydrogen, forming hydrocarbons. These hydrocarbons have the ability to bond with single bonds (alkanes), double bonds (alkenes), or triple bonds (alkynes), and closed-ring compounds, creating many useful products that enhance the quality of our lives.

3–1 SOLUTION CHEMISTRY
THEMATIC FOCUS

The purpose of this section is to introduce students to the defining characteristics of solutions, which are homogeneous mixtures in which molecules are uniformly distributed. The solute and the solvent, the two components of a solution, are defined and distinguished, and electrolytes and nonelectrolytes are contrasted. Students will discover that because matter can exist as a solid, a liquid, or a gas, nine different solution types can be created from these three phases of matter. Factors affecting the rate of solution, solubility, and solution concentration will also be examined.

The themes that can be focused on in this section are systems and interactions, stability, and unity and diversity.

*Systems and interactions: The rate of solution depends on temperature, surface area, and agitation. The solubility of a solute is affected by temperature and pressure.

*Stability: A supersaturated solution is highly unstable.

Unity and diversity: All solutions contain a solute dissolved in a solvent. Solutions can be unsaturated, saturated, or supersaturated, depending on how much solute is dissolved in a given amount of solvent at a given temperature.

PERFORMANCE OBJECTIVES 3–1
1. Define solution.
2. Name the two components of solutions.
3. Compare electrolyte and nonelectrolyte solutions.
4. Compare unsaturated, saturated, and supersaturated solutions.

5. Describe three factors that influence the rate of solution.

SCIENCE TERMS 3–1
solution p. O65
solute p. O66
solvent p. O66
electrolyte p. O67
nonelectrolyte p. O67
solubility p. O69
concentration p. O70
concentrated solution p. O70
dilute solution p. O70
saturated solution p. O70
unsaturated solution p. O71
supersaturated solution p. O71

3–2 ACIDS AND BASES
THEMATIC FOCUS

The physical and chemical properties of acids and bases are described in this section, as well as their effect on indicators. The properties of acids are explained on the basis of the production of hydrogen ions and hydronium ions, and the properties of bases are explained on the basis of the production of hydroxide ions. Acid and base strengths are described, as are common acids and bases.

The themes that can be focused on in this section are energy and unity and diversity.

*Energy: All chemical reactions involved in the formation of acids, bases, salts, and organic compounds either absorb or generate heat.

Unity and diversity: Acids are proton donors; bases are proton acceptors.

PERFORMANCE OBJECTIVES 3–2
1. State the properties of acids and bases.
2. Define and describe the use of indicators.
3. Name several common acids and bases and state their uses.

SCIENCE TERMS 3–2
acid p. O72
base p. O72

3–3 ACIDS AND BASES IN SOLUTION: SALTS
THEMATIC FOCUS

This section will introduce students to the pH scale and its uses. The connection between indicators and pH is discussed, along with the nature of salts and their formation in neutralizing reactions. Precipitation is also explained.

The themes that can be focused on in this section are patterns of change and systems and interactions.

*Patterns of change: The process in which an acid chemically combines with a base is called neutralization. The product of a neutralization reaction are salt and water. The pH of the products is 7.

*Systems and interactions: When acids react chemically with bases, they form salt and water.

PERFORMANCE OBJECTIVES 3–3
1. Explain the nature and use of the pH scale.
2. Describe salt formation in neutralization reactions.
3. Define and explain precipitation.

3–4 CARBON AND ITS COMPOUNDS

THEMATIC FOCUS

In this section students will discover that the vast majority (more than 90 percent) of all chemical compounds contain carbon. Carbon has the ability to bond with many elements, and its role in the creation of many useful products will be examined. Students will be introduced to the concept of a structural formula, which displays the number and position of a molecule's atoms. The structural formulas will enable students to describe isomers of compounds—structures with the same formula but with different shapes and properties.

The themes that can be focused on in this section are scale and structure and unity and diversity.

*Scale and structure: Carbon's ability to combine with itself and with other elements in a variety of ways explains why more than 2 million organic compounds exist.

Unity and diversity: All organic compounds contain carbon. There are many different groups of organic compounds.

PERFORMANCE OBJECTIVES 3–4

1. Describe the physical state of most organic compounds.
2. Diagram the structural formulas of simple organic compounds.
3. Diagram the structural formulas of simple isomers.

SCIENCE TERMS 3–4

3–5 HYDROCARBONS

THEMATIC FOCUS

The purpose of this section is to introduce students to four general classes of hydrocarbons: alkanes, alkenes, alkynes, and aromatic. The special properties of each class will be described. A systematic naming procedure for hydrocarbons is also presented. This organized system teaches students to name hydrocarbons based on the number of continuously bonded carbons within the compound.

The themes that can be focused on in this section are patterns of change and stability.

*Patterns of change: Each alkane, alkene, and alkyne differs from its preceeding series member by CH_2.

*Stability: Saturated hydrocarbons (alkanes) are more stable than unsaturated hydrocarbons (alkenes and alkynes) and thus tend to be less reactive.

PERFORMANCE OBJECTIVES 3–5

1. Define saturated and unsaturated hydrocarbons.
2. Compare the bonding differences between alkanes, alkenes, and alkynes.
3. Explain the increased chemical activity of unsaturated hydrocarbons based on the presence of multiple bonding.

SCIENCE TERMS 3–5

3–6 SUBSTITUTED HYDROCARBONS

THEMATIC FOCUS

This section will explain that when a hydrogen atom is removed from a hydrocarbon and substituted for another type of atom, the resulting compound is called a substituted hydrocarbon. This process results in the creation of alcohols, organic acids, esters, and halogens.

The theme that can be focused on in this section is systems and interactions.

*Systems and interactions: Substituted hydrocarbons are formed when one or more hydrogen atoms in a hydrocarbon chain or ring is replaced by a different atom or group of atoms.

PERFORMANCE OBJECTIVES 3–6

1. Identify the four major groups of substituted hydrocarbons.
2. Describe the formation of the four major groups of substituted hydrocarbons.

SCIENCE TERMS 3–6

Discovery *Learning*

TEACHER DEMONSTRATION MODELING
Observing the Conductivity of Solutions

Set up a low-voltage dry cell, three lengths of insulated copper wire, and a light bulb and fixture. Connect the light-bulb fixture to one of the terminals and attach the second length of wire to the other post of the light-bulb fixture. Attach the third length of wire to the unused terminal of the dry cell. Carefully touch the two free exposed ends of wire together for a moment (handling the insulated parts of the wire), showing that the light bulb lights. Next, fill a beaker with distilled water and place the two free wire ends into it, demonstrating the nonconductivity of water. Repeat the process, using beakers filled with solutions of sugar (a nonelectrolyte) and of sodium chloride (an electrolyte) in water. Ask students to explain what they observe.

CHAPTER 3
Families of Chemical Compounds

INTEGRATING SCIENCE

This physical science chapter provides you with numerous opportunities to integrate other areas of science, as well as other disciplines, into your curriculum. Blue numbered annotations on the student page and integration notes on the teacher wraparound pages alert you to areas of possible integration.

In this chapter you can integrate earth science and geology (p. 64), food science (pp. 64, 72), physical science and mixtures (p. 65), physical science and electrolytes (p. 67), physical science and alloys (p. 68), physical science and heat (p. 69), physical science and antifreeze (p. 71), physical science and chemical equations (p. 77), physical science and organic compounds (p. 78), language arts (p. 79), life science and chemistry of living things (p. 80), earth science and Saturn (p. 84), life science and zoology (p. 89), physical science and halogens (p. 90), physical science and refrigeration (p. 90), and life science and nutrition (p. 91).

SCIENCE, TECHNOLOGY, AND SOCIETY/COOPERATIVE LEARNING

Because more and more chemicals are being used in today's workplace, the number of work-related illnesses, injuries, and deaths related to chemical exposure have increased.

Because some of these illnesses and deaths from hazardous chemicals can occur 20 or more years after exposure in the workplace, employers are hesitant to accept responsibility. Employers argue that they should not have to pay millions of dollars to affected employees because many of the hazards that may have caused their illnesses are also found in today's environment. Employees, on the other hand, argue that employers often hide hazards from the workers and the general public until an accident happens and employers are forced to explain the situation.

During the 1970s, the federal government created the Occupational Safety and Health Administration (OSHA) and the National Institute of Occupational

INTRODUCING CHAPTER 3

DISCOVERY LEARNING

▶ *Activity Book*

Begin teaching the chapter by using the Chapter 3 Discovery Activity from the *Activity Book*. Using this activity, students will explore how various flavors are created using compounds called esters.

USING THE TEXTBOOK

Have students observe the picture on page O62 and its caption on page O63.
• **What do you observe in the picture?** (Various desserts.)
• **How are desserts made?** (Accept all answers. Encourage volunteers to describe the ingredients that might be used to create the desserts shown in the picture.)
• **Do you think chemical compounds are used in foods other than desserts?** (Yes. Have students recall that a compound

Families of Chemical Compounds

Banana, strawberry, pineapple, peach—what's your special flavor? How would you order your favorite ice cream sundae or soda? Certainly not by asking for a scoop of methyl butylacetate! Or a solution of carbonic acid, lactic acid, and ethyl butyrate! Yet that is exactly what you are eating when you enjoy a banana ice cream sundae or a pineapple ice cream soda.

Methyl butylacetate is the banana-flavored compound that makes banana ice cream different from strawberry or vanilla ice cream. And pineapple ice cream owes its characteristic flavor to the presence of ethyl butyrate.

Methyl butylacetate and ethyl butyrate belong to a much larger group of compounds known as organic compounds. Organic compounds are also found in the sugar and cream in the ice cream. But that's not all. In addition to organic compounds, your ice cream sundae or soda contains compounds known as acids, bases, and salts.

In this chapter you will learn about several groups of important compounds—compounds with a variety of surprising and sometimes delicious uses. So sit back and think about a nice big ethyl cinnemate sundae topped with isoamyl salicylate. . . .

Journal *Activity*

You and Your World What do you think of when you hear the words solution, saturated, concentration, base, salt, and neutral? In your journal, write your immediate reaction to each word. When you have finished reading this chapter, see how closely your definitions match the scientific ones.

What delicious examples of the many uses of two important families of chemical compounds!

O ■ 63

---

consists of two or more elements that are chemically combined. Point out that a organic compound is a compound that contains carbon and have students read the chapter introduction on page O63.

• **Why are chemical compounds such as methyl butylacetate and ethyl butyrate used in foods such as desserts and ice cream?** (The chemical compounds help to give certain foods their unique flavors.)

• **Why don't food manufacturing companies use, for example, natural pineapple for food flavoring instead of using chem-** **ical compounds to create an artificial pineapple flavor?** (Accept logical responses. Students might suggest that the cost of using natural flavorings is prohibitive or that the supply of the necessary crops would not be sufficient to meet the demand.)

• **Do you think that artificial chemical compounds should be approved for safety before they are added to foods?** (Yes.)

• **Why do you think so?** (Accept reasonable explanations. Students are likely to cite health concerns.

Safety and Health (NIOSH) to regulate safety and substance use in the workplace. In 1983, the federal government also passed a "right-to-know" law that required employers to provide information to employees about hazardous materials with which they are working.

Using chemicals, businesses have produced new products and processes that are beneficial to society, but many questions remain about the manufacturing process: Should employers be expected to prevent every single hazard? Will the public be willing to pay the additional cost (of the added protection for workers) in higher prices for goods and services? What role should the government play?

Cooperative learning: Using preassigned groups or randomly selected teams, have groups complete the following assignment:

• The federal right-to-know law states that only those employees who work directly with hazardous chemicals must be informed about their dangers. Some states require much more open disclosure—all employees at a facility and even the community in which the facility is located must be made aware of the hazard. What recommendations, if any, would you make to OSHA regarding possible amendments to the federal right-to-know law? Encourage groups to consider the economic, political, social, and environmental impact of their proposed changes.

See Cooperative Learning in the *Teacher's Desk Reference.*

JOURNAL ACTIVITY

You may want to use the Journal Activity as the basis for a class discussion. Have students recall and discuss various situations in which they used these terms or these terms were used. Also discuss why the use of chemicals in foods and other products should be a source of possible concern for a consumer. Students should be instructed to keep their Journal Activity in their portfolio.

3–1 Solution Chemistry

Guide for Reading

Focus on these questions as you read.
▶ What is one important family of compounds?
▶ What is a solution?

Figure 3–1 *The solution process has produced some of nature's loveliest wonders. These stalactites and stalagmites, found in Luray Caverns, Virginia, formed when salts crystallized out of solution. Acids, bases, and salts are found in many common substances, such as these appetizing breads.*

3–1 Solution Chemistry

In Chapter 1 you learned that as a result of chemical bonding (the combining of atoms of elements to form new substances), hundreds of thousands of different substances exist. In Chapter 2 you were introduced to the various chemical reactions by which atoms are rearranged to form new and different substances. In this chapter you will discover how scientists have attempted to bring order to the incredible number of different types of chemical compounds that exist. In other words, you will learn about a system of classifying compounds into families based on their physical and chemical properties.

One of the most important and abundant families of chemical compounds is the family of acids, bases, and salts. This family may already be familiar to you. Have you ever heard of acetic acid in vinegar? Magnesium hydroxide in milk of magnesia? Or sodium hydrogen carbonate in baking soda? Acetic acid is an acid; magnesium hydroxide is a base; and sodium hydrogen carbonate is a salt. See—you do know something about this family!

In order to better understand acids, bases, and salts, it will be helpful to take a few steps back and look at an important process that produces these compounds. The process involves making solutions.

What Is a Solution?

What happens when a lump of sugar is dropped into a glass of lemonade? What takes place when carbon dioxide gas is bubbled through water? And where do mothballs go when they disappear? The answer to these questions is the same: The sugar, gaseous carbon dioxide, and mothballs all dissolve in the substances in which they are mixed.

Careful examination of each of these mixtures—even under a microscope—will not reveal molecules of sugar in lemonade, carbon dioxide in water, or naphthalene in the air. But the sweet taste of lemonade tells you the sugar is there. The "fizziness" of soda water indicates the presence of carbon dioxide. And the smell of mothballs is evidence of naphthalene. In each of these mixtures, the molecules of the substance have become evenly distributed among the molecules of the other substance. The mixtures are uniform (the same) throughout. Each mixture is a **solution.**

Figure 3–2 *Solutions abound in nature. Blood is a solution. It contains fats, salts, sugars, and proteins dissolved in water. The dragon that guards the Forbidden City in China is made of bronze, which is a type of solution known as an alloy. A solution of limestone rock and water formed a cavern, which then collapsed and formed a sinkhole . What solutions can you identify in this scene of Glacier Park, Montana?* ①

O ■ 65

Use the breads shown in Figure 3–1 to integrate food science concepts into your lesson.

CONTENT DEVELOPMENT

Write the term *homogeneous* on the chalkboard. Point out that a homogeneous mixture is a mixture that appears to be the same throughout. Another way to describe homogeneous is "well-mixed." Explain to students that a solution is a special kind of homogeneous mixture that actually has the same makeup throughout the mixture. The substances in a solution are not chemically combined and can be separated. Solutions can be solids dissolved in liquids, liquids dissolved in liquids, gases dissolved in liquids, or gases dissolved in gases. Stress that the key to knowing that a substance is a solution is the term *dissolved.*

● ● ● ● **Integration** ● ● ● ●

Use the examples of lemonade and mothballs to integrate concepts of mixtures into your lesson.

INDEPENDENT PRACTICE

📀 Media and Technology

You might want to show the video/videodisc called Elements, Compounds, and Mixtures to reinforce the concept of the impact of chemistry on society and to extend understanding of elements, compounds, and mixtures. The video explores how the purity of a substance affects its properties, the classification of mixtures as heterogeneous or homogeneous, the use of physical properties and changes to separate mixtures, and the use of atomic structure to account for the uniqueness of compounds. After viewing the video, have students work in small groups and classify a variety of substances as elements, compounds, or mixtures.

ACTIVITY
DOING
EMULSIFYING ACTION

Skills: Making observations, relating concepts

Materials: oil, vinegar, container with cover, egg yolk

In this activity students should notice that the oil and the vinegar do not dissolve in each other. Though these liquids appear to dissolve upon shaking, they soon separate after the shaking stops. Adding the egg yolk makes the two liquids dissolve upon shaking. Students should determine that the oil and vinegar are a suspension; the oil, vinegar, and egg yolk is a colloid; and a common name for the end product is mayonnaise.

ACTIVITY DOING

Emulsifying Action

1. To 10 mL of oil, add an equal amount of vinegar. Do the two liquids dissolve in each other?

2. Shake the liquids and observe what immediately happens. What happens after several minutes?

3. Now add an egg yolk to the mixture and shake again. Describe what happens.

The egg yolk acts as an emulsifying agent, changing one type of mixture to another.

4. Find out what the words colloid and suspension mean.

In this activity, which is the colloid? The suspension?

What is a common name for your end product?

A solution is a mixture in which one substance is dissolved in another substance. Different parts of a solution are identical. That is what is meant by the words uniform throughout. The molecules making up a solution are too small to be seen and do not settle when the solution is allowed to stand. A solution, then, is a "well-mixed" mixture.

All solutions have several basic properties. Let's go back to the glass of sweetened lemonade to discover just what these properties are. **A solution consists of two parts: One part is the substance being dissolved, and the other part is the substance doing the dissolving.**

The substance that is dissolved is called the **solute** (SAHL-yoot). The substance that does the dissolving is called the **solvent** (SAHL-vuhnt). The solvent is sometimes called the dissolving medium. In the sweetened lemonade, the solute is the sugar and the solvent is the lemonade. Even without the sugar, the lemonade is a solution. It is made of water and lemon juice. What are the solutes and solvents in the other two examples of solutions you have just read about? ❶

The most common solutions are those in which the solvent is a liquid. The solute can be a solid, a liquid, or a gas. The most common solvent is water.

Figure 3–3 *A solution consists of a solute and a solvent. The most common solutions are those in which the solvent is a liquid. Here you see the three types of solutions that can be formed from a liquid solvent and a solid, a liquid, and a gas solute. What is the most common liquid solvent?* ❷

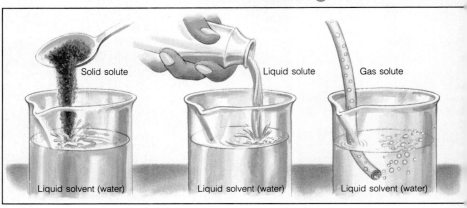

Solid solute — Liquid solute — Gas solute

Liquid solvent (water) — Liquid solvent (water) — Liquid solvent (water)

3–1 (continued)

CONTENT DEVELOPMENT

Contrast solutions and other kinds of mixtures, taking sufficient time to make it clear that only solutions are truly mixtures at the "molecular level." In other words, solutions are mixtures in which molecules of the dissolved substances are not clumped together but are surrounded by and interspersed with molecules of the substance in which they are dissolved. (Have students recall the description "well-mixed.")

Also make clear the distinction between solute and solvent, pointing out that a solvent is the substance that does the dissolving in a solution, whereas a solute is the substance that is dissolved in a solution. Provide examples of solutions and, in each case, ask students:

• **Which substance is the solute, and which is the solvent?** (Answers will vary, depending on the particular examples chosen.)

GUIDED PRACTICE

Skills Development
Skill: Applying Concepts

Ask students to prepare a heterogeneous mixture of fine sand in water and a homogeneous solution of salt water.

Working in groups, ask students to devise a different way in which the solutes of each could be separated. (Students might suggest that the sand could be separated from the water by means of filtration, using a funnel and filter paper, and that

the salt could be separated from the water by means of boiling or evaporation.)

Provide groups with filter paper, a heat source, and other equipment necessary to separate the solutes by filtering and boiling or evaporation. If students will be using a heat source, remind them to work sensibly and ask groups to describe and account for what occurs in each process.

CONTENT DEVELOPMENT

Make a clear distinction between electrolytes and nonelectrolytes. Point out

Figure 3–4 *A liquid solution appears clear. Rocks known as geodes form when a solvent evaporates, leaving behind deposits of solute.*

When alcohol is the solvent, the solution is called a tincture. Have you ever heard of tincture of iodine? It is an antiseptic used to treat minor cuts and scratches. What is the solute in this solution? ❸

The particles in a solution are individual atoms, ions, or molecules. Because the particles are so small, they do not scatter light that passes through the solution. A liquid solution appears clear.

Most solutions cannot easily be separated by simple physical means such as filtering. However, a physical change such as evaporation or boiling can separate the parts of many solutions. If salt water is boiled, the water will change from a liquid to a gas, leaving behind particles of salt.

Another important property of a solution is its ability or inability to conduct an electric current. An electric current is a flow of electrons. In order for electrons to flow through a solution, ions must be present. As you learned in Chapter 1, ions are charged atoms. A solution that contains ions is a good conductor of electricity. A solution that does not contain ions is a nonconductor.

Pure water is a poor conductor of electric current because it does not contain ions. If a solute such as potassium chloride (KCl) is added to the water, however, the resulting solution is a good conductor. Substances whose water solutions conduct an electric current are called **electrolytes** (ee-LEHK-troh-lights). Potassium chloride, sodium chloride, and silver nitrate are examples of electrolytes. Most electrolytes are ionic compounds.

Substances whose water solutions do not conduct an electric current are called **nonelectrolytes.** A solution of sugar and water does not conduct an electric current. Sugar is a nonelectrolyte, as are alcohol and benzene. Many covalent compounds are nonelectrolytes because they do not form ions in solution.

Making Solutions

Solutions abound in nature. The oceans, the atmosphere, even Earth's interior are solutions. Your body contains a number of vital solutions. Every solution has a particular solute and solvent.

O ■ 67

that an important property of a solution is its ability or inability to conduct an electric current.

Some solutions do not conduct electric current because the solutions lack the presence of ions. These solutions are known as nonelectrolyte solutions. Many covalent compounds are nonelectrolytes because they do not form ions in solution. (Have students recall that covalent bonds involve the sharing of electrons between atoms.)

Solutions in which ions are present are known as electrolyte solutions. These solutions have the ability to conduct electricity. Sodium chloride and silver nitrate are examples of electrolytes. Most electrolytes are ionic compounds. (Have students recall that ionic bonding involves the transfer of electrons between atoms.)

● ● ● ● **Integration** ● ● ● ●

Use the example of solutions that conduct electric current to integrate concepts of electrolytes into your lesson.

GAS SOLUBILITY
AND TEMPERATURE

The inverse dependence of gas solubility on the temperature of the liquid solvent accounts in part for the fact that tropical waters, which are at relatively high temperatures, do not support large numbers of certain types of organisms that do not thrive under conditions in which dissolved oxygen is low in concentration.

TYPES OF SOLUTIONS

Solute	Solvent	Example
Gas	Gas	Air (oxygen in nitrogen)
Gas	Liquid	Soda water (carbon dioxide in water)
Gas	Solid	Charcoal gas mask (poisonous gases on carbon)
Liquid	Gas	Humid air (water in air)
Liquid	Liquid	Antifreeze (ethylene glycol in water)
Liquid	Solid	Dental filling (mercury in silver)
Solid	Gas	Soot in air (carbon in air)
Solid	Liquid	Ocean water (salt in water)
Solid	Solid	Gold jewelry (copper in gold)

Figure 3–5 *Nine different types of solutions can be made from the three phases of matter. What are solutions of solids dissolved in solids called?* ❶

TYPES OF SOLUTIONS Matter can exist as a solid, a liquid, or a gas. From these three phases of matter, nine different types of solutions can be made. Figure 3–5 shows these types of solutions.

The most common solutions are liquid solutions, or solutions in which the solvent is a liquid. The solute can be a solid, a liquid, or a gas. Two liquids that dissolve in each other are said to be miscible (MIHS-uh-buhl). Water and alcohol are miscible. Do you think oil and water are miscible? Solutions of solids dissolved in solids are called alloys. Most alloys are made of metals. Some common alloys include brass, bronze, solder, stainless steel, and wrought iron. You might want to find out exactly what the solute and solvent are in each of these alloys.

RATE OF SOLUTION Suppose you wanted to dissolve some sugar in a glass of water as quickly as possible. What might you do? If your answer included stirring the solution, using granulated sugar, or heating the water, you are on the right track.

Normally, the movement of solute molecules away from the solid solute and throughout the solvent occurs rather slowly. Stirring or shaking the solution helps to move solute particles away from the solid solute faster. More molecules of the solute are brought in contact with the solvent sooner, so the solute dissolves at a faster rate.

Figure 3–6 *Scientists have developed a model to describe a probable solution process. First, solute particles separate from the surface of the solid solute. Then, the solute molecules enter the liquid surrounding the solid solute. Finally, the solute molecules are attracted to the solvent molecules.*

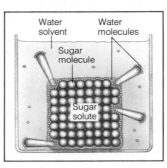

Water solvent

Water molecules

Sugar molecule

Sugar solute

68 ■ O

3–1 (continued)

CONTENT DEVELOPMENT

Briefly review the phases of matter and demonstrate that there are nine solute-solvent combinations by phase. Direct students' attention to Figure 3–5, which illustrates these nine types of solutions, and ask students to name other examples of the various types.

Also, explain the factors affecting the rate of dissolving, pointing out the connection between temperature and average molecular velocity.

● ● ● ● **Integration** ● ● ● ●

Use the example of solids dissolved in solids to integrate concepts of alloys into your lesson.

Use the description of how temperature change affects the rate of solution to integrate heat concepts into your lesson.

INDEPENDENT PRACTICE

▶ *Activity Book*

Students who need practice on the concept of different types of solutions should complete the chapter activity Solutions Everywhere. Using this activity, students will examine their environment and provide examples of the nine different types of solutions.

CONTENT DEVELOPMENT

You might consider having students carry out a simple experiment to determine the effects of various factors on the rate of solution. Have groups measure the time it takes for identical small sugar cubes to dissolve completely in beakers containing equal amounts of water. The sugar cube in the first beaker, which should contain cold water and which should not be stirred, will dissolve very slowly. In fact, some sugar will probably remain undissolved after several hours. The second beaker should contain water that is also cold but stirred continuously after the cube is put in. The third beaker should contain cold water that is not to be stirred; the sugar cube placed in it should first be pulverized into fine powder, using a mortar and pestle. The fourth beaker,

Solution action occurs only at the surface of the solid solute. So if the surface area of the solute is increased, the rate of solution is increased. When the solid solute is ground into a fine powder, more solute molecules are in contact with the solvent. Finely powdered solids dissolve much faster than do large lumps or crystals of the same substance.

If heat is applied to a solution, the molecules move faster and farther apart. As a result, the dissolving action is speeded up.

SOLUBILITY From experience, you know that table salt and sugar readily dissolve in water. These compounds are described as being very soluble or having a high degree of **solubility** (sahl-yoo-BIHL-uh-tee) in water. The solubility of a solute is a measure of how much of that solute can be dissolved in a given amount of solvent under certain conditions. Although a large amount of table salt dissolves in water, only a small amount of table salt dissolves in alcohol. So although the solubility of salt in water is high, the solubility of salt in alcohol is rather low. The solubility of a solute depends on the nature of the solute and of the solvent. Solubility is usually described in terms of the mass of the solute that can be dissolved in a definite amount of solvent at a specific temperature.

Two main factors affect the solubility of a solute. These factors are temperature and pressure. Usually, an increase in the temperature of a solution increases the solubility of a solid in a liquid. Just think how much more sugar dissolves in hot tea than in cold.

ACTIVITY DISCOVERING

Solubility of a Gas in a Liquid

1. Remove the cap from a bottle of soda.
2. Immediately fit the opening of a balloon over the top of the bottle. Shake the bottle several times. Note any changes in the balloon.
3. Heat the bottle of soda very gently by placing it in a pan of hot water. Note any further changes in the balloon.
- What two conditions of solubility are being tested?
- What general statement about the solubility of a gas in a liquid can you now make?

Figure 3–7 *The solubility of a gas solute in a liquid solvent depends on both the pressure and the temperature. When the cap is removed from the bottle, the solubility of the gas decreases. If the bottle is cold, the decrease in solubility is very small. If the bottle is warm, the decrease is obvious.*

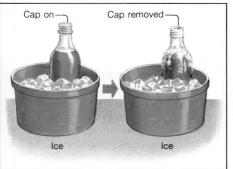

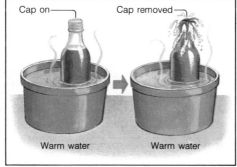

O ■ 69

ACTIVITY DISCOVERING

SOLUBILITY OF A GAS IN A LIQUID

Discovery Learning

Skills: Making observations, making generalizations

Materials: bottle of soda, balloon, pan

This activity will help students relate the concept of solubility to pressure and temperature. Students should infer that the two conditions of solubility being tested are pressure and temperature, and they should generalize that raising the temperature of a solution decreases the solubility of a gas in a liquid.

---

into which a whole sugar cube is to be placed, should first be heated to near its boiling point. (**CAUTION:** *Students should be told about the safe and sensible handling of the heat source and the hot materials.*)
• **What effect did the various factors have on the rate of solution?** (Stirring the water, powdering the sugar, and heating the water all increased the rate of dissolving.)

GUIDED PRACTICE
▶ *Laboratory Manual*
Skills Development
Skill: Making observations
At this point you may want to have students complete the Chapter 3 Laboratory Investigation in the *Laboratory Manual* called Relating Solubility and Temperature. In this investigation students will determine how the temperature of a solution affects a solute's ability to dissolve.

ENRICHMENT
▶ *Activity Book*
Students will be challenged by the Chapter 3 activity in the *Activity Book* called Solutions and Temperature. In this activity students will explore the importance of temperature in dissolving substances in a solution.

Figure 3–8 *This sight may be a common one to you — and now you know enough about solubility to explain it. What is happening here?* ②

ACTIVITY
DISCOVERING

Sudsing Power

1. Separately add 0 mL, 10 mL, 20mL, and 40 mL of salt to four 1-liter jars each containing 500 mL of water. Stir until the salt dissolves.

2. Add 10 mL of laundry soap to each jar.

3. Cap each jar and vigorously shake for 20 seconds.

■ How did the different concentrations of salt affect the sudsing action of the laundry soap?

■ What effect does water that contains large amounts of dissolved minerals have on washing clothes and on bathing? Do you know the name of this type of water?

The situation for a gas dissolved in a liquid is just the opposite. Raising the temperature of a solution decreases the solubility of a gas in a liquid. Perhaps you have observed this fact without actually realizing it. Have you ever let a glass of soda get warm? If so, what did you notice? The soda goes flat, or loses its fizz. The fizz in soda is carbon dioxide gas dissolved in soda water. As the temperature of the solution increases, the solubility of the carbon dioxide decreases. The gas comes out of solution, leaving the soda flat. Why do you think boiled water tastes flat?

For both solids and liquids dissolved in liquids, increases and decreases in pressure have practically no effect on solubility. For gases dissolved in liquids, however, an increase in pressure increases solubility, and a decrease in pressure decreases solubility. A bottle of soda fizzes when the cap is removed because molecules of carbon dioxide gas escape from solution as the pressure is decreased. The solubility of the carbon dioxide gas has been decreased by a decrease in pressure.

CONCENTRATION The **concentration** of a solution refers to the amount of solute dissolved in a certain amount of solvent. A solution in which a lot of solute is dissolved in a solvent is called a **concentrated solution.** A solution in which there is little solute dissolved in a solvent is called a **dilute solution.** The terms concentrated and dilute are not precise, however. They do not indicate exactly how much solute and solvent are present.

Using the concept of solubility, the concentration of a solution can be expressed in another way. A solution can be described as saturated, unsaturated, or supersaturated. To understand these descriptions, remember that solubility measures the maximum amount of solute that can be dissolved in a given amount of solvent.

A **saturated solution** is a solution that contains all the solute it can possibly hold at a given temperature. In a saturated solution, no more solute can be dissolved at that temperature. If more solute is added to a saturated solution, it will settle undissolved to the bottom of the solution. In describing a saturated solution, the temperature must always be given. Can you explain why? ③

An **unsaturated solution** is a solution that contains less solute than it can possibly hold at a given temperature. In an unsaturated solution, more solute can be dissolved.

Under special conditions, a solution can be made to hold more solute than is normal for that temperature. Such a solution is a **supersaturated solution.** A supersaturated solution is unstable. If a single crystal of solute is added to a supersaturated solution, the excess solute comes out of solution and settles to the bottom. Only enough solute to make the solution saturated remains dissolved.

SPECIAL PROPERTIES Why is salt spread on roads and walkways that are icy? Why is salt added to cooking water? Why is a substance known as ethylene glycol added to the cooling systems of cars? The answers to these questions have to do with two special properties of solutions.

Experiments show that when a solute is dissolved in a liquid solvent, the freezing point of the solvent is lowered. The lowering of the freezing point is called freezing point depression. The addition of solute molecules interferes with the phase change of solvent molecules. So the solution can exist in the liquid phase at a lower temperature than can the pure solvent. Ethylene glycol, commonly known as ❶ antifreeze, is added to automobile cooling systems to lower the freezing point of water.

The addition of a solute to a pure liquid solvent also raises the boiling point of the solvent. This increase is called boiling point elevation. In this case,

Figure 3–9 *A saturated solution contains all the solute it can possibly hold at a given temperature. So any additional solute will fall to the bottom of the solution (left). Crystals of sugar are growing on a string placed in a supersaturated solution of sugar and water (center). Adding a small crystal of solute starts the crystallizing action instantly. The formation of these stalactites is also the result of solute crystallization from a supersaturated solution (right).*

INTEGRATION
MATHEMATICS

Metal alloys have many practical uses. Some useful metallic materials with which students are familiar are actually alloys. For example, steel is an alloy of iron, carbon, and small amounts of other elements. Bronze is a copper-tin alloy. Brass is a copper-zinc alloy. Silver used in jewelry is a silver-copper alloy, and yellow gold used in jewelry is a gold-copper alloy.

● ● ● ● **Integration** ● ● ● ●

Use the dependence of automobile cooling systems on ethylene glycol to integrate antifreeze concepts into your lesson.

ENRICHMENT

Solution concentrations can be expressed in a number of different ways, including molarity (the number of moles of solute per liter of solution) and molality (the number of moles of solute per kilogram of solvent.)

GUIDED PRACTICE

Skills Development
Skill: Relating cause and effect

To help students better understand solution concentrations, prepare a saturated water solution of sodium acetate before class, using hot water as the solvent. Decant the saturated solution so that no undissolved crystals remain in it. Allow the solution to cool slowly; then add a single crystal of sodium acetate to the now-supersaturated solution. Ask:

• **What do you observe?** (Rapid precipitation occurs.)

• **How can you explain this observation?** (The solution must have been supersaturated.)

REINFORCEMENT/RETEACHING

▶ *Activity Book*

Students who need practice on the concept of solubility should be provided with the Chapter 3 activity called Dissolving Household Substances. In this activity students will discover how quickly various solid substances typically found in a home dissolve in water.

3-2 Acids and Bases

3-1 (continued)

Figure 3–10 *Salt is spread on icy surfaces because it causes the ice to melt at a lower temperature. Rock salt is used in an ice cream maker to lower the freezing point of the ice that surrounds the container of ice cream. This means that the ice melts at and remains at a temperature below 0°C. This, in turn, allows for more efficient cooling of the ice cream mixture and the eventual freezing of the ice cream.*

the addition of solute molecules interferes with the rapid evaporation, or boiling, of the solvent molecules. So the solution can exist in the liquid phase at a higher temperature than can the pure solvent. When salt is added to cooking water, the water boils at a higher temperature. Although it takes a longer time to heat the water to boiling, it takes a shorter time to cook food in that water.

3-1 Section Review

1. What is one important family of compounds? What process produces these compounds?
2. Compare a solute and a solvent.
3. What are three properties of a solution?
4. Describe three ways in which you can increase the rate at which a solid dissolves in a liquid.
5. Compare a saturated, unsaturated, and supersaturated solution.

Connection—*You and Your World*
6. Why is the antifreeze ethylene gylcol used in automobile radiators in warm weather?

3-2 Acids and Bases

If you look in your medicine cabinet and refrigerator and on your kitchen shelves, you will find examples of groups of compounds known as **acids** and **bases.** Acids are found in aspirin, vitamin C, and eyewash. Fruits such as oranges, grapes, lemons, grapefruits, and apples contain acids. Milk and tea contain acids, as do pickles, vinegar, and carbonated drinks. Bases are found in products such as lye, milk of magnesia, deodorants, ammonia, and soaps.

72 ■ O

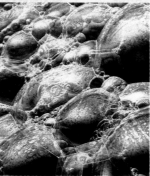

Figure 3–11 *Many of the fruits you eat contain acids. Acids are also used in the manufacture of synthetic fibers. The soap that produced these bubbles was manufactured using bases.*

Acids and bases also play an important role in the life processes that take place in your body. Many industrial processes use acids and bases. The manufacture of a wide variety of products involves the use of acids and bases.

Properties of Acids

As a class of compounds, all acids have certain physical and chemical properties when dissolved in water. One of the physical properties all acids share is sour taste. Lemons taste sour because they contain citric acid. Vinegar contains acetic acid. However, you should never use taste to identify a chemical substance. You should use other, safer properties.

Acids affect the color of indicators. Indicators are compounds that show a definite color change when mixed with an acid or a base. Litmus paper, a common indicator, changes from blue to red in an acid solution. Another indicator, phenolphthalein (fee-nohl-THAL-een), is colorless in an acid solution.

Acids react with active metals to form hydrogen gas and a metal compound. This reaction wears away, or corrodes, the metal and produces a residue. For example, sulfuric acid in a car battery often corrodes the terminals and leaves a residue.

Another property of acids can be identified by looking at the list of common acids in Figure 3–12 on page 74. What do all these acids have in common? Acids contain hydrogen. When dissolved

ACTIVITY DOING

Acid-Base Testing

1. Ask your teacher for several strips of red litmus paper and blue litmus paper.
2. Test these substances with the litmus paper to determine if they are an acid or a base: orange juice, milk, tea, coffee, soda, vinegar, ammonia cleaner, milk of magnesia, saliva.

Report the results of your tests to your classmates by preparing a poster in which you classify each substance.

O ■ 73

ACTIVITY DOING
ACID-BASE TESTING

Skills: Making observations, making comparisons

Materials: red litmus paper, blue litmus paper, orange juice, milk, tea, coffee, soda, vinegar, ammonia cleaner, milk of magnesia, saliva

Check each poster to ensure the accuracy of each classification.

TEACHING STRATEGY 3–2

FOCUS/MOTIVATION

Without using the terms *acid* and *base,* use the chalkboard to list several common acidic food substances, such as lemon juice, lime juice, and vinegar, and several common base substances, such as antacids and laxatives. Draw a large circle around the acids and another large circle around the bases, then ask students what properties they think these circled groups of substances have in common.

Point out that these groups of substances are acids and bases and that students will learn about their properties and uses in this section of the textbook.

CONTENT DEVELOPMENT

Explain that when acids are mixed with water, they produce the positive hydrogen ion, H^+. This is a distinguishing property of acids.

solute it can possibly hold at a given temperature, an unsaturated solution contains less solute than it can possibly hold at a given temperature, and a supersaturated solution contains more solute than is normal for that temperature.

6. The addition of ethylene glycol interferes with the rapid evaporation, or boiling, of the cooling-system solvent, causing the solution to have the potential to exist in the liquid phase at a higher temperature, decreasing the potential of overheating.

REINFORCEMENT/RETEACHING

Monitor students' responses to the Section Review questions. If students appear to have difficulty with any of the questions, review the appropriate material in the section.

CLOSURE

▶ *Review and Reinforcement Guide*

At this point have students complete Section 3–1 in the *Review and Reinforcement Guide.*

There are a number of different definitions for acids. Substances that produce H^+ ions and that have the properties ordinarily associated with acids are called Arrhenius acids. Substances that donate protons in reactions are called Bronsted-Lowry acids. Substances that accept electron pairs in reactions are called Lewis acids.

3–2 (continued)

INDEPENDENT PRACTICE

Media and Technology

Use the transparency called Acids, Bases, and pH Values to help develop the concept of acids and bases.

GUIDED PRACTICE

▶ *Laboratory Manual*

Skills Development

Skill: Making observations

At this point you may want to have students complete the Chapter 3 Laboratory Investigation in the *Laboratory Manual* called Testing Unknown Substances Using Acid-Base Indicators. In this investigation students will discover how to determine whether a substance is an acid or a base.

CONTENT DEVELOPMENT

Before the discovery was made that acids contain hydrogen, a number of other elements were proposed as the one element common to acids. For example, the eighteenth-century French scientist Lavoisier proposed that all acids contain oxygen. The first scientist who proposed that all acids contain hydrogen was French chemist Claude Louis Berthollet (1748–1822), who made this proposal in 1798.

GUIDED PRACTICE

▶ *Laboratory Manual*

Skills Development

Skill: Making observations

At this point you may want to have students complete the Chapter 3 Laborato-

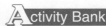

Activity Bank

In a Jam, p.156

COMMON ACIDS	
Name and Formula	**Uses**
Strong	
Hydrochloric HCl	Pickling steel; cleaning bricks and metals; digesting food
Sulfuric H_2SO_4	Manufacturing paints, plastics, fertilizers; dehydrating agent
Nitric HNO_3	Removing tarnish; making explosives (TNT); making fertilizers
Weak	
Carbonic H_2CO_3	Carbonating beverages
Boric H_3BO_3	Washing eyes
Phosphoric H_3PO_4	Making fertilizers and detergents
Acetic $HC_2H_3O_2$	Making cellulose acetate used in fibers and films
Citric $H_3C_6H_5O_7$	Making soft drinks

in water, acids ionize to produce positive hydrogen ions (H^+). A hydrogen ion is a proton. So acids are often defined as proton donors.

The hydrogen ion, or proton, produced by an acid is quickly surrounded by a water molecule. The attraction between the hydrogen ion (H^+) and the water molecule (H_2O) results in the formation of a hydronium ion, H_3O^+.

The definition of an acid as a proton donor helps to explain why all hydrogen-containing compounds are not acids. Table sugar contains 22 hydrogen atoms, but it is not an acid. When dissolved in water, table sugar does not produce H^+ ions. Table sugar is not a proton donor. So it does not turn litmus paper red or phenolphthalein colorless.

Common Acids

The three most common acids in industry and in the laboratory are sulfuric acid (H_2SO_4), nitric acid (HNO_3), and hydrochloric acid (HCl). These three acids are strong acids. That means they ionize to a high degree in water and produce hydrogen ions. The presence of hydrogen ions makes strong acids good electrolytes. (Remember, electrolytes are good conductors of electricity.)

Acetic acid ($HC_2H_3O_2$), carbonic acid (H_2CO_3), and boric acid (H_3BO_3) are weak acids. They do not ionize to a high degree in water, so they produce few hydrogen ions. Weak acids are poor electrolytes. Figure 3–12 lists the name, formula, and uses of some common acids. Remember, handle any acid—weak or strong—with care!

Properties of Bases

When dissolved in water, all bases share certain physical and chemical properties. Bases usually taste bitter and are slippery to the touch. However, bases can be poisonous and corrosive. So you should never use taste or touch to identify bases.

Figure 3–12 *The name, formula, and uses of some common acids are given in this chart. What is the difference between a strong acid and a weak acid?* ❶

ry Investigation in the *Laboratory Manual* called Acids and Bases From Home. In this investigation students will investigate how the acidity or alkalinity of common household products can be determined.

REINFORCEMENT/RETEACHING

Students sometimes confuse the terms *concentrated* and *strong*. Point out that a concentrated acid suspension contains many dissolved acid molecules per unit volume. A strong acid is an acid whose molecules tend to ionize to a high degree. As a result, strong acids can exist in concentrated or in dilute solutions, as can weak acids.

GUIDED PRACTICE

▶ *Laboratory Manual*

Skills Development

Skill: Making observations

At this point you may want to have students complete the Chapter 3 Laboratory Investigation in the *Laboratory Manual* called Testing Commercial Antacids. In

Bases turn litmus paper from red to blue and phenolphthalein to bright pink. Bases emulsify, or dissolve, fats and oils. They do this by reacting with the fat or oil to form a soap. The base ammonium hydroxide is used as a household cleaner because it "cuts" grease. The strong base sodium hydroxide, or lye, is used to clean clogged drains.

All bases contain the hydroxide ion, OH⁻. When dissolved in water, bases produce this ion. Because a hydroxide ion (OH⁻) can combine with a hydrogen ion (H⁺) and form water, a base is often defined as a proton (H⁺) acceptor.

Common Bases

Strong bases dissolve readily in water to produce large numbers of ions. So strong bases are good electrolytes. Examples of strong bases include potassium hydroxide (KOH), sodium hydroxide (NaOH), and calcium hydroxide ($Ca(OH)_2$).

Weak bases do not produce large numbers of ions when dissolved in water. So weak bases are poor electrolytes. Ammonium hydroxide (NH_4OH) and aluminum hydroxide ($Al(OH)_3$) are weak bases. See Figure 3–13.

COMMON BASES

Name and Formula	Uses
Strong	
Sodium hydroxide NaOH	Making soap; drain cleaner
Potassium hydroxide KOH	Making soft soap; battery electrolyte
Calcium hydroxide Ca(OH)₂	Leather production; making plaster
Magnesium hydroxide Mg(OH)₂	Laxative; antacid
Weak	
Ammonium hydroxide NH₄OH	Household cleaner
Aluminum hydroxide Al(OH)₃	Antacid; deodorant

3–2 Section Review

1. What are three important properties of acids? Of bases?
2. Why are acids called proton donors?
3. Why are bases called proton acceptors?
4. If an electric conductivity setup were placed in the following solutions, would the light be bright or dim: HCl, HNO_3, H_3BO_3, $HC_2H_3O_2$, NH_4OH, KOH, NaOH, $Al(OH)_3$?

Connection—*Laboratory Safety*

5. How could you safely determine whether an unknown solution is an acid or a base?

ECOLOGY NOTE
LIMESTONE CAVES

Carbon dioxide in the air dissolves in water to form carbonic acid. The carbonic acid then reacts with limestone rock ($CaCO_3$), causing it to decompose. The result is the formation of limestone caves.

INDEPENDENT PRACTICE

Section Review 3–2

1. Acids: turn blue litmus paper red; react with active metals to produce hydrogen gas and metal compounds; contain hydrogen, and when dissolved in water produce positive hydrogen ions; taste sour. Bases: taste bitter; slippery to the touch; turn litmus paper from red to blue and phenolphthalein bright pink; emulsify fats and oils; contain hydroxide ion.

2. When dissolved in water, acids ionize to produce positive hydrogen ions, or protons.

3. When dissolved in water, the hydroxide ion of the base combines with a hydrogen ion, or proton.

4. Bright; bright; dim; dim; dim; bright; bright; dim.

5. Use litmus paper or phenolphthalein.

REINFORCEMENT/RETEACHING

Review students' responses to the Section Review questions. Reteach any material that is still unclear, based on students' responses.

CLOSURE

▶ *Review and Reinforcement Guide*

Students may now complete Section 3–2 in the *Review and Reinforcement Guide*.

this investigation students will explore whether or not commercial stomach antacids neutralize acid.

REINFORCEMENT/RETEACHING

▶ *Activity Book*

Students who need practice on the concept of pH values should be provided with the Chapter 3 activity called Acid, Base, or Neutral Substance? In this activity students will predict the pH of various substances.

GUIDED PRACTICE

Skills Development

Skill: *Making comparisons*

Have students design and create charts that illustrate the differences between acids and bases in terms of structure and properties. They may also add illustrations of their own making or cut from magazines that show some uses for acids and bases.

3-3 Acids and Bases in Solution: Salts

MULTICULTURAL OPPORTUNITY 3-3

Did you know that salt had the power to bring down an empire? During the fight for independence with England, India used salt as an important example of nonviolent civil disobedience. Previously, salt was taxed by England. Under the leadership of Gandhi, Indian people marched to the sea and distilled their own salt, in direct violation of the tax laws.

ESL STRATEGY 3-3

Have students complete the following paragraph:

Two forms of solutions are _____ and _____ ; however, solutions can also be _____. To determine how much _____ a solution contains, the pH scale is used. It is a series of numbers from _____ to _____. Its neutral point is _____, which indicates that the solution is neither a (an) _____ _____ or a(an) _____ _____. Strong acids have _____ pH numbers; strong bases have _____ pH numbers. _____ are formed from the positive ion of a(an) _____ and the negative _____ of a(an) _____ _____; therefore, they are _____ substances.

Activity Bank

Soda Fountain, Activity Book, p. O157. This activity can be used for ESL and/or Cooperative Learning.

Guide for Reading

Focus on these questions as you read.
▶ What is pH?
▶ How are salts formed?

3-3 Acids and Bases in Solution: Salts

As you have just learned, solutions can be acidic or basic. Solutions can also be neutral. To measure the acidity of a solution, the **pH scale** is used. **The pH of a solution is a measure of the hydronium ion (H_3O^+) concentration.** Remember, the hydronium ion is formed by the attraction between a hydrogen ion (H^+) from an acid and a water molecule (H_2O). So the pH of a solution indicates how acidic the solution is.

The pH scale is a series of numbers from 0 to 14. The middle of the scale—7—is the neutral point. A neutral solution has a pH of 7. It is neither an acid nor a base. Water is a neutral liquid.

A solution with a pH below 7 is an acid. Strong acids have low pH numbers. Would hydrochloric acid have a pH closer to 2 or to 6? ❶

A solution with a pH above 7 is a base. Strong bases have high pH numbers. What would be the pH of NaOH? ❷

Figure 3-14 Does it surprise you to learn that many of the substances you use every day contain acids and bases? Which fruit is most acidic? What cleaner is most basic? ❸

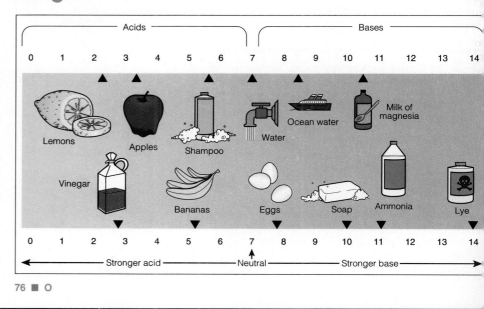

76 ■ O

TEACHING STRATEGY 3-3

FOCUS/MOTIVATION

Set up test tubes containing solutions of varying concentrations of H^+ and varying concentrations of OH^-. Dampen pieces of pH paper with the various solutions. You will obtain the colors red, orange, yellow, green, and blue. Explain to students that the paper had been soaked in a mixture of indicators and is sensitive to different concentrations of H^+ and OH^- ions. Point out that a scale expressing these ranges of concentration in a convenient way would be very useful.

CONTENT DEVELOPMENT

Stress that the pH scale measures the concentration of H^+ ions in solutions. The pH of a solution alone is not a direct indication of acid strength or even of acid concentration. For example, a strong acid at fairly low concentration can produce many H^+ ions and can thus have a low pH. With sufficient dilution, however, H^+ concentration falls and pH rises, even for strong acids. A weak acid at high concentration can have a low pH because, al-

though the tendency of each molecule to ionize (that is, the strength) is low, there are many molecules present to ionize.

Also point out the number range of the pH scale and make sure students understand that solutions with high H^+ ion concentration have a low pH.

● ● ● ● **Integration** ● ● ● ●

Use the neutralization-reaction formula to integrate concepts of chemical equations into your lesson.

Determining Solution pH

The pH of a solution can be determined by using an indicator. You already know about two indicators: litmus paper and phenolphthalein. Other indictors include pH paper, methyl orange, and bromthymol blue. Each indicator shows a specific color change as the pH of a solution changes.

Common household materials can be used as indicators. Red-cabbage juice covers the entire pH range. Grape juice is bright pink in the presence of an acid and bright yellow in the presence of a base. Even tea can be an indicator. Have you ever noticed the color of tea change when you add lemon juice? For accurate pH measurements, a pH meter is used.

Formation of Salts

When acids react chemically with bases, they form a class of compounds called salts. A **salt** is a compound formed from the positive ion of a base and the negative ion of an acid. A salt is a neutral substance.

The reaction of an acid with a base produces a salt and water. The reaction is called **neutralization** (noo-truhl-ih-ZAY-shuhn). In neutralization, the properties of the acid and the base are lost as two neutral substances—a salt and water—are formed.

The reaction of HCl with NaOH is a neutralization reaction. The positive hydrogen ion from the acid combines with the negative hydroxide ion from the base. This produces water. The remaining positive ion of the base combines with the remaining negative ion of the acid to form a salt.

$$H^+Cl^- + Na^+OH^- \longrightarrow H_2O + NaCl$$ 1

Many of the salts formed by a neutralization reaction are insoluble in water—that is, they do not dissolve in water. They crystallize out of solution and remain in the solid phase. An insoluble substance that crystallizes out of solution is called a precipitate (pree-SIHP-uh-tayt). The process by which a precipitate forms is called precipitation. An example of a precipitate is magnesium carbonate. Snow, rain, sleet, and hail are considered forms of precipitation because they fall out of solution. Out of what solution do they precipitate? 4

ACTIVITY DOING

A Homemade Indicator

1. Shred two leaves of red cabbage and boil them in some water until the liquid gets very dark. **CAUTION:** *Wear safety goggles and use extreme care .*

2. When the liquid and the shreds have cooled, squeeze all the purple juice you can from the shreds. Pour the liquid into a container.

3. Into 5 separate containers, pour about 3 to 4 mL of the following substances, one in each container: shampoo, grapefruit juice, clear soft drink, milk, ammonia cleaner.

4. Add about 2 mL of the cabbage liquid. Stir and observe any color changes.

Based on the fact that ammonia is a base, identify the other substances as either acids or bases.

ANNOTATION KEY

Answers

1 Closer to 2. (Applying concepts)
2 About 14. (Applying concepts)
3 Lemon; lye. (Interpreting charts)
4 Air. (Inferring)

Integration

1 Physical Science: Chemical Equations. See *Matter: Building Block of the Universe*, Chapter 2.

ACTIVITY DOING

A HOMEMADE INDICATOR

Skills: Making comparisons, identifying patterns

Materials: red cabbage, safety goggles, 5 separate containers, shampoo, grapefruit juice, clear soft drink, milk, ammonia cleaner, graduated cylinder

This activity will show students that substances in everyday life can be used in a scientific manner. Check students' observations for each of the substances listed in the activity, or substitute substances of your choice.

INDEPENDENT PRACTICE

Media and Technology

You might want to show the Video/Videodisc called Acids, Bases, and Salts to reinforce an appreciation of the significance of these compounds to living things and to extend understanding of the operational definitions and characteristics of these related compounds. The video explores the classification of acids, bases, and salts as electrolytes; a comparison of strong and weak acids and bases; the use of the pH concept to compare the acidity or basicity of solutions; and an identification of changes in a neutralization reaction. After viewing the video, have students work in groups to predict, then test, the acidity or basicity of various substances.

INDEPENDENT PRACTICE

▶ *Product Testing Activity*

Have students perform the product test on antacids from the Product Testing Activity worksheets. Ask students to relate the antacid's effectiveness in neutralizing acids.

REINFORCEMENT/RETEACHING

▶ *Activity Book*

Students who need practice on the concept of different types of salts should be provided with the Chapter 3 activity called Names and Uses of Salts. In this activity students will learn about various widely used salts.

3–4 Carbon and Its Compounds

MULTICULTURAL OPPORTUNITY 3–4

Have students research the life and work of Dr. Lloyd Ferguson. Dr. Ferguson was a chemistry teacher for more than 35 years. At Howard University, he was a role model for African-American students wishing to pursue careers in chemistry. Among his contributions was the authorship of the text *Modern Structural Theory of Organic Chemistry.*

ESL STRATEGY 3–4

Ask students these questions:
1. What element is found in more than 90 percent of all known compounds?
2. What are these compounds called?
3. Why are there so many of these compounds?

While they are learning the properties of organic compounds, have students rearrange the following list in alphabetical order: gases, liquids, low-melting solids, strong odors, low boiling points, insoluble, nonelectrolytes, low melting points.

3–3 (continued)

INDEPENDENT PRACTICE

▶ *Activity Book*

Students who need practice on the concept of chemical compounds should complete the chapter activity Acid, Base, or Salt? Using this activity, students will use the formula for a chemical compound to determine facts about that compound.

GUIDED PRACTICE

Skills Development

Skills: Making comparisons, applying concepts

At this point have students complete the in-text Chapter 3 Laboratory Investigation: Acids, Bases, and Salts. In this investigation students will identify various properties of acids and bases and will discover what happens when acids react with bases.

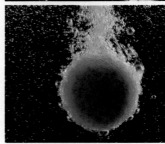

Figure 3–15 *To control dangerous acid spills, firefighters spray a blanket of base over the acid. The result is a neutral salt and water. These two substances are also produced when a person uses an antacid to relieve an upset stomach. What reaction is taking place in both of these situations?* ❶

A neutralization reaction is a double-replacement reaction—and a very important one, too! For a dangerous acid can be combined with a dangerous base to form a harmless salt and neutral water.

3–3 Section Review

1. What is pH? Describe the pH scale.
2. What is the pH of an acid? A base? Water?
3. How can the pH of a solution be determined?
4. How does a salt form? What is a salt?
5. What is neutralization?

Critical Thinking—*Applying Concepts*
6. Use an equation to show the neutralization reaction between H_2SO_4 and NaOH.

Guide for Reading

Focus on these questions as you read.

▶ *What are organic compounds?*
▶ *What are some properties of organic compounds?*

3–4 Carbon and Its Compounds

Do you know what such familiar substances as sugar, plastic, paper, and gasoline have in common? They all contain the element carbon. Carbon is present in more than 2 million known compounds, and this number is rapidly increasing. Approximately 100,000 new carbon compounds are being isolated or synthesized every year. In fact, more than 90 percent of all known compounds contain carbon!

❶ **Carbon compounds form an important family of chemical compounds known as organic compounds.** The word organic means coming from life. Because carbon-containing compounds are present in all living things, scientists once believed that **organic compounds** could be produced only by living organisms. Living things were thought to have a mysterious "vital force" that was responsible for creating

INDEPENDENT PRACTICE

Section Review 3–3

1. The pH of a solution is a measure of the concentration of hydronium ions. The pH scale is a scale of numbers from 0 to 14 in which 7 is neutral, less than 7 is acidic, and greater than 7 is basic.
2. The pH of an acid is between 0 and 7; the pH of a base is between 7 and 14; the pH of water, if it is pure, is 7.
3. It can be determined by using an indicator such as litmus paper, phenolphthalein, pH paper, or methyl orange. A pH meter can be used for more accurate measurement.
4. A salt forms when acids react chemically with bases; a salt is a compound formed from the positive ion of a base and the negative ion of an acid.
5. Neutralization is a chemical reaction in which an acid and a base combine to produce a salt and water.
6. $H_2SO_4 + 2NaOH = Na_2SO_4 + 2H_2O$.

Figure 3–16 *The element carbon is present in more than 2 million known compounds. Here you see three different forms of the pure element: diamond, graphite, and coal.*

carbon compounds. It was believed that the force could not be duplicated in the laboratory.

In 1828, the German chemist Friedrich Wöhler produced an organic compound called urea from two inorganic substances. Urea is a waste product produced by the human body. It was not long before chemists accepted the idea that organic compounds could be prepared from materials that were never part of a living organism. What is common to all organic compounds is not that they originated in living things but that they all contain the element carbon. Today, the majority of organic compounds are synthesized in laboratories.

There are some carbon compounds that are not considered organic compounds. Calcium carbonate, carbon dioxide, and carbon monoxide are considered inorganic (not organic) compounds.

The Bonding of Carbon

Carbon's ability to combine with itself and with other elements explains why there are so many carbon compounds. Carbon atoms form covalent bonds with other carbon atoms.

The simplest bond involves 2 carbon atoms. The most complex involves thousands of carbon atoms. The carbon atoms can form long straight chains, branched chains, single rings, or rings joined together.

The bonds between carbon atoms can be single covalent bonds, double covalent bonds, or triple covalent bonds. In a single bond, one pair of electrons is shared between 2 carbon atoms. In a double

ACTIVITY
DOING

Octane Rating

1. Find out what the octane rating of gasoline means.
2. Go to a local gas station and find out the octane ratings of the different grades of gasoline being sold. Compare the prices of the different grades. Ask the station attendant to describe how each grade of gasoline performs.
3. Present your findings to your class. ❷

O ■ 79

ACTIVITY
DOING
OCTANE RATING

Skills: Relating concepts, making comparisons

This activity gives students an opportunity to relate organic chemistry to the real-life process of petroleum refining and combustion. Students should discover that combustion in an automobile engine should not occur before the spark plugs ignite the mixture; otherwise, engine "knocking" will occur. The octane-number scale uses values from 0 to 100 to rate the antiknock characteristics of a particular gasoline.

Integration: Use this Activity to integrate language arts skills into your science lesson.

REINFORCEMENT/RETEACHING

Monitor students' responses to the Section Review questions. If students appear to have difficulty with any of the questions, review the appropriate material in the section.

CLOSURE

▶ *Review and Reinforcement Guide*

At this point have students complete Section 3–3 in the *Review and Reinforcement Guide.*

TEACHING STRATEGY 3–4

FOCUS/MOTIVATION

Explain to students that a fuel such as gasoline, which is combusted to provide energy for transportation, is a mixture of various hydrocarbons. A hydrocarbon consists of molecules of carbon bonded to carbon and to hydrogen. This type of bonding results in nonpolar molecules.
• **Around self-service gasoline pumps, you may notice an occasional spill from the pump. Why do these spills evaporate so easily?** (Hydrocarbons are nonpolar molecules with little attraction between molecules. The change from a liquid into a gaseous state will therefore require only small amounts of energy.)

● ● ● ● **Integration** ● ● ● ●

Use the explanation of carbon compounds to integrate concepts of organic compounds into your lesson.

CAR WAXES

Most hydrocarbons are nonpolar. Car waxes are largely high-molecular-weight nonpolar hydrocarbons. In addition to improving appearance, the hydrocarbons in wax keep a nonpolar layer between the metal of the automobile and the polar water in the environment.

Figure 3–17 *Because of carbon's bonding ability, a great variety of organic compounds exists. These include synthetic rubber and plastic for firefighters' equipment, paints and dyes, and the substances of which all living things are made.*

bond, two pairs of electrons are shared between 2 carbon atoms. See Figure 3–18. How many pairs of electrons are shared in a triple bond? ❶

Carbon atoms also bond with many other elements. These elements include oxygen, hydrogen, members of the nitrogen family, and members of Family 17. The simplest organic compounds contain just carbon and hydrogen. Because there are so many compounds of carbon and hydrogen, they form a class of organic compounds all their own. You will soon read about this class of compounds.

A great variety of organic compounds exists because the same atoms that bond together to form one compound may be arranged in several other ways in several other compounds. Each different arrangement of atoms represents a separate organic compound.

Properties of Organic Compounds

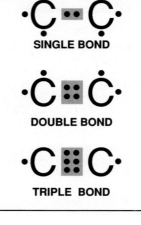

SINGLE BOND

DOUBLE BOND

TRIPLE BOND

Organic compounds usually exist as gases, liquids, or low-melting-point solids. Organic liquids generally have strong odors and low boiling points. Organic liquids do not conduct an electric current. What is the name for a substance whose solution does not conduct electricity? Organic compounds generally ❷ do not dissolve in water. Oil, which is a mixture of organic compounds, floats on water because the two liquids are insoluble.

Figure 3–18 *In a single bond, one pair of electrons is shared. In a double bond, two pairs of electrons are shared. How many pairs of electrons are shared in a triple bond?* ❸

3–4 (continued)

CONTENT DEVELOPMENT

Point out that carbon has the ability to bond with itself and other elements and that this ability is the reason why there are many organic compounds. Have students note the examples of organic compounds, or compounds containing carbon, pictured in Figure 3–17.

Direct students' attention to Figure 3–18. Explain that in a single bond, one

pair of electrons is shared between 2 carbon atoms.

• **How many electron pairs are shared in a double bond?** (Two pairs.)

Have students note that in a triple bond, three pairs of electrons are shared.

Direct students' attention to Figure 3–19. Explain that the organic compound ethane has 2 carbon atoms and 4 hydrogen atoms per molecule. Because carbon has 4 valence electrons, each electron will bond covalently with an electron of another atom, in this case hydrogen, to pro-

duce a stable outermost level. Point out that this ethane molecule is an example of a structural formula.

• **In the structural formula for methane, one molecule contains how many carbon atoms?** (1.)

• **How many hydrogen atoms?** (4.)

• **In the structural formula for propane, one molecule contains how many carbon atoms?** (3.)

• **How many hydrogen atoms?** (8.)

Structural Formulas

A molecular formula for a compound indicates what elements make up that compound and how many atoms of each element are present in a molecule. For example, the molecular formula for the organic compound ethane is C_2H_6. In every molecule of ethane, there are 2 carbon atoms and 6 hydrogen atoms.

What a molecular formula does not indicate about a molecule of a compound is how the different atoms are arranged. To do this, a **structural formula** is used. A structural formula shows the kind, number, and arrangement of atoms in a molecule. You can think of a structural formula as being a model of a molecule.

Figure 3–19 shows the structural formula for ethane and two other organic compounds: methane and propane. Note that in a structural formula, a dash (–) is used to represent the pair of shared electrons forming a covalent bond. In writing structural formulas, it is important that you remember the electron arrangement in a carbon atom.

Carbon has 4 valence electrons, or 4 electrons in its outermost energy level. Each electron will form a covalent bond with an electron of another atom to produce a stable outermost level containing 8 electrons. So when structural formulas are written, there can be no dangling bonds—no dangling dashes.

METHANE
CH_4

ETHANE
C_2H_6

PROPANE
C_3H_8

Figure 3–19 *Methane, ethane, and propane are the first three members of a series of hydrocarbons known as the alkanes. Note that each carbon atom is surrounded by four dashes, corresponding to four pairs of shared electrons.*

Isomers

Compounds with the same molecular formula but different structures are called **isomers.** Figure 3–20 on page 82 shows two isomers of butane, C_4H_{10}. Notice that one isomer is a straight chain and

ACTIVITY DOING

Isomers

For this activity you will need a stack of blank index cards, a felt-tipped marker, and a package of straws or pipe cleaners.

1. Take half of the index cards and label one side of each card with the letter C.

2. Take the rest of the index cards and label each card with the letter H.

3. Using the straws or pipe cleaners to represent bonds, see how many isomers you can make for the formula C_6H_{14}. Remember that each carbon atom must be bonded to four other atoms.

ACTIVITY DOING

ISOMERS

Skills: Applying concepts, relating facts

Materials: blank index cards, felt-tipped marker, straws or pipe cleaners

This activity will help students grasp the concept of isomers. Students will discover that many different isomers are possible for the structural formula C_6H_{14}. Check students' isomers carefully to make sure each carbon atom is bonded to four other atoms.

GUIDED PRACTICE

Skills Development

Skill: Identifying patterns

Remind students that carbon must have four bonds and hydrogen must have one.
• **Predict the formula of the hydrocarbon octane.** (C_8H_{18}.)

Have a volunteer diagram the unbranched structural formula of octane on the chalkboard. Have another volunteer diagram two branched isomers of octane on the chalkboard. Ask the remainder of the class to examine the diagrams and discuss their accuracy.

● ● ● ● **Integration** ● ● ● ●

Use Figure 3–17 to integrate concepts of cells into your lesson.

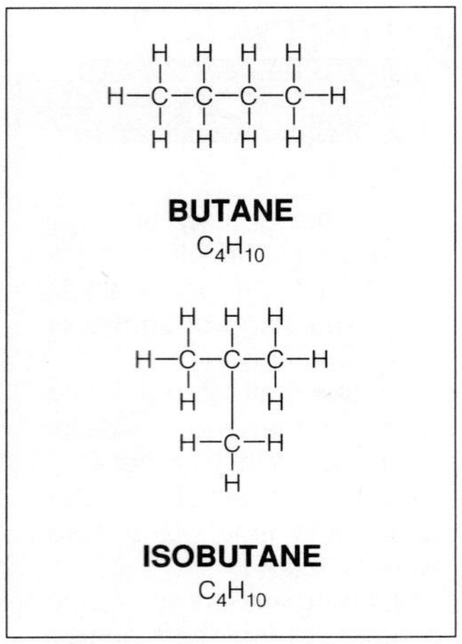

Figure 3–20 *Butane has two isomers: normal butane and isobutane. Which isomer is a branched chain?* **①**

BUTANE
C_4H_{10}

ISOBUTANE
C_4H_{10}

3–4 (continued)

REINFORCEMENT/RETEACHING

When dealing with isomers for the first time, students may confuse two identical structures. For example, the middle pentane isomers in Figure 3–21 could be flopped right to left, which is actually the same isomer shown in the textbook, even though the attached carbon group appears to be attached to the third carbon instead of the second. Actually, the two are mirror images, not structural isomers.

ENRICHMENT

Although the structures shown in Figures 3–19 and 3–20 indicate that the hydrocarbons are in a straight line, the actual molecular shape is twisted. For example, the propane molecule in Figure 3–19 appears to have the three carbons arranged along a 180-degree bond angle. The actual measured angles would show an angle of 109 degrees from one carbon to the next, resulting in a "jagged" structure. The hydrogen atoms are also 109 degrees away from each carbon atom. This three-dimensional twisting gives the molecules several contact points to cause some weak attractions.

The 109-degree angle results because of an electron rearrangement with the carbon atom. The orbitals in the *s* and *p* sublevels rearrange to form new orbitals for bonding. These new orbitals—four for carbon—are 109 degrees apart (the shape of a tetrahedron).

the other isomer is a branched chain. In a branched chain, all the carbon atoms are not in a straight line. This difference in structure will account for any differences in the physical and chemical properties of these two compounds.

Figure 3–21 shows three isomers of pentane, C_5H_{12}. This time there is one straight chain and two branched chains. To see the difference between the two branched chains, count the number of carbon atoms in the straight-chain portion of each molecule. How many are there in each branched isomer? What do you think happens to the number of possible isomers as the number of carbon atoms in a molecule increases? The compound whose formula is $C_{15}H_{32}$ could have more than 400 isomers!

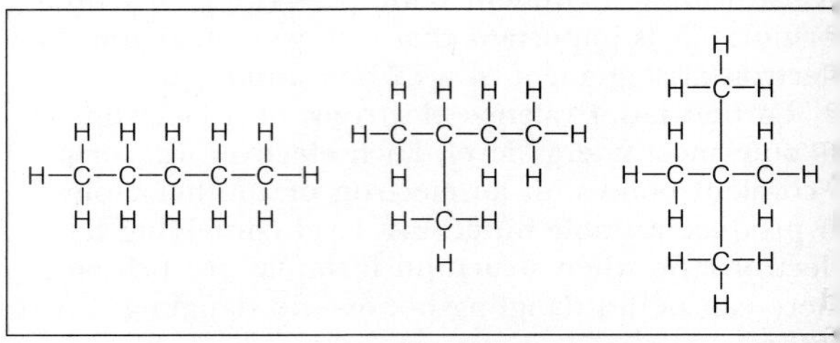

Figure 3–21 *As the number of carbon atoms increases, the number of isomers increases. What alkane is shown here? How do its three isomers differ?* **④**

3–4 Section Review

1. What are organic compounds?
2. What four factors account for the abundance of carbon compounds?
3. What are three properties of organic compounds?

Critical Thinking—*Applying Concepts*

4. Could two compounds have the same structural formula but different molecular formulas?

INDEPENDENT PRACTICE

Section Review 3–4

1. Organic compounds are compounds that contain the element carbon.

2. Carbon atoms form covalent bonds with other carbon atoms. The bonds between carbon atoms can be single, double, or triple covalent bonds. Carbon atoms bond with many other elements. The same atoms may be arranged in several different ways.

3. Organic compounds typically exist as gases, liquids, and low-melting solids, and they do not dissolve in water. Liquid organic compounds typically have a strong odor, are nonelectrolytic, and have a low boiling point. Solid organic compounds typically have a low melting point.

4. No, in order to have the same structural formula, both compounds would have to have the same kind and same number of atoms. This is the only information a molecular formula gives.

3–5 Hydrocarbons

Have you ever heard of a butane lighter, seen a propane torch, or noticed a sign at a service station advertising "high octane" gasoline? Butane, propane, and octane are members of a large group of organic compounds known as **hydrocarbons.** A hydrocarbon contains only hydrogen and carbon.

Hydrocarbons can be classified as saturated or unsaturated depending on the type of bonds between carbon atoms. In **saturated hydrocarbons,** all the bonds between carbon atoms are single covalent bonds. In **unsaturated hydrocarbons,** one or more of the bonds between carbon atoms is a double covalent or triple covalent bond.

Alkanes

The **alkanes** are straight-chain or branched-chain hydrocarbons in which all the bonds between carbon atoms are single covalent bonds. Alkanes are saturated hydrocarbons. All the hydrocarbons that are alkanes belong to the alkane series. The simplest member of the alkane series is methane, CH_4. Methane consists of 1 carbon atom surrounded by 4 hydrogen atoms. Why are there 4 hydrogen atoms? ⑤

The next simplest alkane is ethane, C_2H_6. How does the formula for ethane differ from the formula for methane? After ethane, the next member of the

Figure 3–22 *Petroleum is one of the most abundant sources of hydrocarbons. The hydrocarbons in petroleum range from 1-carbon molecules to more than 50-carbon molecules. Petroleum, which does not mix with water, is highly flammable.*

3–5 Hydrocarbons

MULTICULTURAL OPPORTUNITY 3–5

Students might think of science as being rather systematic and sometimes boring. Relate this legend to show the importance of imagination in science. August Kekule, the German chemist, was struggling with an understanding of the structure of carbon compounds. One night as he was sitting in front of his fireplace dozing off, he began to dream about a group of six monkeys, each monkey grabbing the tail of the monkey in front of it. This dream led him to develop the idea of a ringed structure for some carbon compounds.

ESL STRATEGY 3–5

To reinforce some of the terms in this section, have students match the terms in column A with the definitions in column B. Then have volunteers read their answers aloud.

A
• Hydrocarbons
• Saturated hydrocarbons
• Unsaturated hydrocarbons
• Alkanes
• Prefix
• A type of saturated hydrocarbon
• Suffix

B
• Part of a compound's name that indicates to which series it belongs
• An organic compound that contains only hydrogen and carbon
• All the bonds between carbon atoms are single covalent bonds in this type of hydrocarbon
• One or more of the bonds between carbon atoms is a double covalent or triple bond
• A type of saturated hydrocarbon
• Part of a compound's name that indicates the number of carbon atoms it contains

REINFORCEMENT/RETEACHING

Review students' responses to the Section Review questions. Reteach any material that is still unclear, based on students' responses.

CLOSURE

▶ *Review and Reinforcement Guide*
Students may now complete Section 3–4 in the *Review and Reinforcement Guide.*

TEACHING STRATEGY 3–5

FOCUS/MOTIVATION

Bring samples or labels of several different brands of margarines, cooking oils, and shortenings to class. Ask students to find and compare the amounts of saturated and unsaturated fats in each product. These metric values can be written on the chalkboard and used to explain the differences in bonding between saturated and unsaturated hydrocarbons.

Figure 3-23 *Common alkanes include methane, of which the planet Saturn's atmosphere is composed; propane, which is burned to provide heat for hot-air balloons; and butane, which is the fuel in most lighters. What is the general formula for the alkanes?* ❶

ALKANE SERIES	
Name	**Formula**
Methane	CH_4
Ethane	C_2H_6
Propane	C_3H_8
Butane	C_4H_{10}
Pentane	C_5H_{12}
Hexane	C_6H_{14}
Heptane	C_7H_{16}
Octane	C_8H_{18}
Nonane	C_9H_{20}
Decane	$C_{10}H_{22}$

84 ■ O

alkane series is propane, C_3H_8. Can you begin to see a pattern to the formulas for each successive alkane? Ethane has 1 more carbon atom and 2 more hydrogen atoms than methane does. Propane has 1 more carbon atom and 2 more hydrogen atoms than ethane does. Each member of the alkane series is formed by adding 1 carbon atom and 2 hydrogen atoms to the previous compound.

The pattern that exists for the alkanes can be used to determine the formula for any member of the series. Because each alkane differs from the preceding member of the series by the group CH_2, a general formula for the alkanes can be written. That general formula is C_nH_{2n+2}. The letter n is the number of carbon atoms in the alkane. What would be the formula for a 15-carbon hydrocarbon? For a 30-carbon hydrocarbon? ❷

Naming Hydrocarbons

Figure 3-24 shows the first ten members of the alkane series. Look at the names of the compounds. How is each name the same? How is each different?

Often in organic chemistry the names of the compounds in the same series will have the same ending, or suffix. Thus, the members of the alkane series all end with the suffix *-ane,* the same ending as in the series name. The first part of each name, or the prefix, indicates the number of carbon atoms

Figure 3-24 *This chart shows the names and formulas for the first ten members of the alkane series. What does the prefix hex- mean?* ❹

resent in the compound. The prefix *meth-* indicates carbon atom. The prefix *eth-*, 2 carbon atoms, and e prefix *prop-*, 3. According to Figure 3–24, how any carbon atoms are indicated by the prefix *pent-*? ow many carbon atoms are in octane? As you study her hydrocarbon series, you will see that these pre-xes are used again and again. So it will be useful r you to become familiar with the prefixes that ean 1 to 10 carbon atoms.

lkenes

Hydrocarbons in which at least one pair of rbon atoms is joined by a double covalent bond e called **alkenes.** Alkenes are unsaturated hydro-rbons. The first member of the alkene series is hene, C_2H_4. The next member of the alkene series propene, C_3H_6.

Figure 3–25 shows the first seven members of the kene series. What do you notice about the name of ch compound? 6

As you look at the formulas for the alkenes, you ll again see a pattern in the number of carbon d hydrogen atoms added to each successive com-ound. The pattern is the addition of 1 carbon atom d 2 hydrogen atoms. The general formula for the kenes is C_nH_{2n}. The letter n is the number of car-on atoms in the compound. What is the formula r an alkene with 12 carbons? With 20 carbons? 8

In general, alkenes are more reactive than al-anes because a double bond is more easily broken an a single bond. So alkenes can react chemically adding other atoms directly to their molecules.

lkynes

Hydrocarbons in which at least one pair of rbon atoms is joined by a triple covalent bond e called **alkynes.** Alkynes are unsaturated ydrocarbons. The simplest alkyne is ethyne, C_2H_2, hich is commonly known as acetylene. Perhaps you ave heard of acetylene torches, which are used in elding.

ALKENE SERIES	
Name	**Formula**
Ethene	C_2H_4
Propene	C_3H_6
Butene	C_4H_8
Pentene	C_5H_{10}
Hexene	C_6H_{12}
Heptene	C_7H_{14}
Octene	C_8H_{16}

Figure 3–25 *This chart shows the names and formulas of the first seven members of the alkene series. What would a 9-carbon alkene be called?* 7

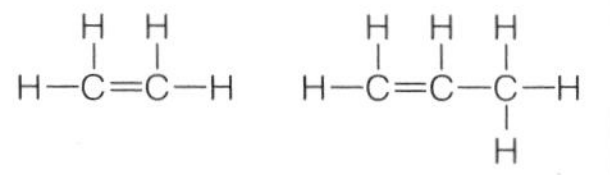

ETHENE C_2H_4 **PROPENE** C_3H_6

gure 3–26 *The first two members of the alkene series are hene and propene. What kind of bonds do the alkenes have?* 9

ANNOTATION KEY

Answers

1 C_nH_{2n+2} (Relating facts)
2 $C_{15}H_{32}$; $C_{30}H_{62}$. (Applying concepts)
3 Ending *-ane;* beginning of name. (Relating facts)
4 Six. (Relating facts)
5 Five; eight. (Applying facts)
6 All end in *-ene,* the series name. (Applying facts)
7 Nonene. (Inferring)
8 $C_{12}H_{24}$; $C_{20}H_{40}$. (Applying concepts)
9 Double covalent. (Interpreting diagrams)

Integration

1 Earth Science: Saturn. See *Exploring the Universe,* Chapter 2.

ENRICHMENT

Changing an unsaturated substance into a saturated compound is an important industrial process called hydrogenation. The process does not require energy to break the hydrogen molecule into individual atoms to react with the multiple bond in the unsaturated compound. The process, however, can be greatly speeded up with a platinum catalyst. This can be a way to change unsaturated oils into saturated shortenings.

INDEPENDENT PRACTICE

▶ *Product Testing Activity*

Have students perform the product test on cereals from the Product Testing Activity worksheets. Ask students to relate the nutrient levels of the cereals to their studies of chemical compounds.

GUIDED PRACTICE

Skills Development

Skill: Interpreting patterns

To help students remember the relationship between multiple bonding and resulting formulas, have them practice recognizing the types of compounds represented by the following:

C_2H_4 (unsaturated: alkene)
CH_4 (saturated: alkane)
C_5H_8 (unsaturated: alkyne)
C_7H_{14} (unsaturated: alkene)

GUIDED PRACTICE

▶ *Laboratory Manual*

Skills Development

Skill: Relating concepts

At this point you may want to have students complete the Chapter 3 Laboratory Investigation in the *Laboratory Manual* called How Much Vitamin C Is in Fruit Juice? In this investigation students will discover how the amount of Vitamin C in fruit juice can be determined.

ALKYNE SERIES	
Name	**Formula**
Ethyne	C_2H_2
Propyne	C_3H_4
Butyne	C_4H_6
Pentyne	C_5H_8
Hexyne	C_6H_{10}

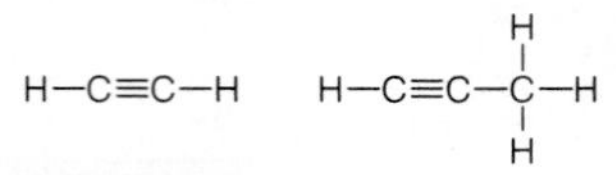

H—C≡C—H

ACETYLENE (ETHYNE)
C_2H_2

H—C≡C—C—H (with H's on the right carbon)

PROPYNE
C_3H_4

Figure 3–27 *This chart shows the names and formulas for the first five members of the alkyne series. What is the general formula for this series? The simplest alkynes are ethyne and propyne. What is the common name for ethyne?* ❶

The first five members of the alkyne series are listed in Figure 3–27. Here again, each successive member of the alkyne series differs by the addition of 1 carbon atom and 2 hydrogen atoms. The general formula for the alkynes is C_nH_{2n-2}.

The alkynes are even more reactive than the alkenes. Very little energy is needed to break a triple bond. Like the alkenes, alkynes can react chemically by adding other atoms directly to their molecules.

Aromatic Hydrocarbons

All the hydrocarbons you have just learned about—the alkanes, alkenes, and alkynes—are either straight-chain or branched-chain molecules. But this is not the only structure a hydrocarbon can have. Some hydrocarbons are in the shape of rings. Probably the best-known class of hydrocarbons in the shape of rings is the aromatic hydrocarbons. The name of this class comes from the fact that aromatic hydrocarbons share a common physical property. These compounds have strong and often pleasant odors (or aromas).

Figure 3–28 *The simplest aromatic hydrocarbon is benzene. Benzene is used in the manufacture of dyes. What is the basic structure of aromatic hydrocarbons?* ❷

3–5 (continued)

CONTENT DEVELOPMENT

Atoms that are double or triple bonded to one another are held closely together. This is because the positive nuclei are more attracted to the electron-rich area that a multiple bond provides. Other reacting atoms may also be attracted to the electron-rich area.

Benzene is usually presented as a structure with alternating double and single bonds. Refer students to Figure 3–29. If the structure actually had some double bonds and some single bonds, however, the shape would be distorted because the double-bonded carbon atoms would be closer together than the single-bonded atoms are.

INDEPENDENT PRACTICE

▶ *Activity Book*

Students who need practice on the concept of hydrocarbons should complete the chapter activity Digestion. Using this activity, students will discover the role of enzymes in the digestive process.

ENRICHMENT

Benzene rings form the basis for many types of compounds. Moth crystals are sold as paradichlorobenzene. Assign interested students to find the formula of this moth-killing compound ($C_6H_4Cl_2$), and another moth-crystal product, naphthalene ($C_{10}H_8$).

INDEPENDENT PRACTICE

Section Review 3–5

1. Hydrocarbons are organic compounds that contain only hydrogen and carbon; hydrocarbons are classified as saturated or unsaturated, depending on the type of bonds between the carbon atoms.

2. Alkanes, alkenes, and alkynes.

3. Saturated compounds have only single bonds. Alkanes are saturated hydrocarbons. Unsaturated compounds have at least one double or triple bond. Alkenes and alkynes are unsaturated hydrocarbons.

4. Aromatic hydrocarbons consist of a ring of 6 carbon atoms joined by alternating single and double covalent bonds.

5. No, it has a hexagonal shape.

The basic structure of an aromatic hydrocarbon is a ring of 6 carbon atoms joined by alternating single and double covalent bonds. This means that within the 6-carbon ring, there are 3 carbon-to-carbon double bonds. The simplest aromatic hydrocarbon is called benzene, C_6H_6. Figure 3–29 shows the structural formula for benzene. Chemists often abbreviate this formula by drawing a hexagon with a circle in the center.

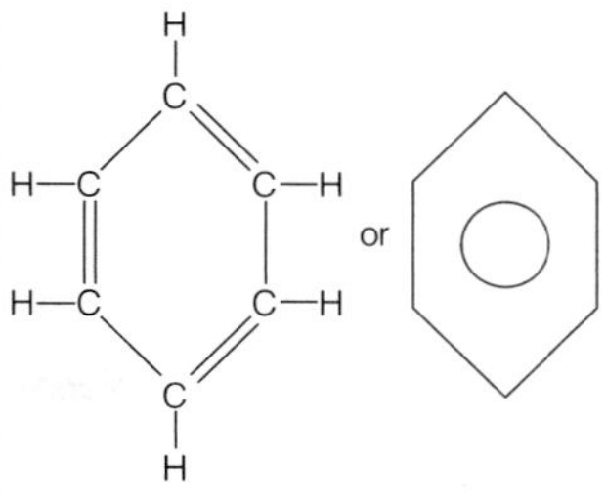

BENZENE
C_6H_6

Figure 3–29 *The structural formula for benzene shows 6 carbon atoms joined by alternating single and double covalent bonds.*

3–5 Section Review

1. What are hydrocarbons? How are hydrocarbons classified?
2. Name three series of hydrocarbons.
3. What is meant by saturated and unsaturated hydrocarbons? Classify each hydrocarbon series according to these definitions.
4. What are aromatic hydrocarbons?

Critical Thinking—*Applying Concepts*
5. Can C_6H_6 be a straight-chain hydrocarbon?

3–6 Substituted Hydrocarbons

Hydrocarbons are but one of several groups of organic compounds. Hydrocarbons contain carbon and hydrogen atoms only. But as you have learned, carbon atoms form bonds with many other elements. So many different groups of organic compounds exist. **The important groups of organic compounds include alcohols, organic acids, esters, and halogen derivatives.** These compounds are called **substituted hydrocarbons.** A substituted hydrocarbon is formed when one or more hydrogen atoms in a hydrocarbon chain or ring is replaced by a different atom or group of atoms.

Guide for Reading

Focus on this question as you read.

▶ *What are some important groups of substituted hydrocarbons?*

3–6 Substituted Hydrocarbons

MULTICULTURAL OPPORTUNITY 3–6

Have students investigate DDT as an example of a substituted hydrocarbon. Although banned in the United States because of environmental damage, it is still used in some developing countries for the prevention of malaria. This can lead to a discussion of the trade-offs between the needs of the environment and the needs of public health.

ESL STRATEGY 3–6

Ask an English-speaking student to volunteer as tutor in helping ESL students prepare answers to the following questions.
1. What is the difference between a hydrocarbon and a substituted hydrocarbon? Name four organic compound groups called substituted hydrocarbons.
2. To create the name for an alcohol, what suffix is used? Name three types of alcohol.
3. In organic acids, which group has replaced hydrogen atoms, as the hydroxyl group has in alcohol?
4. Where are organic acids found? To create their names, what suffix is used? Name three types of organic acids.
5. What is created when an alcohol and an organic acid are chemically combined? Describe this compound in a short paragraph.

REINFORCEMENT/RETEACHING

Monitor students' responses to the Section Review questions. If students appear to have difficulty with any of the questions, review the appropriate material in the section.

CLOSURE

▶ *Review and Reinforcement Guide*
At this point have students complete Section 3–5 in the *Review and Reinforcement Guide.*

TEACHING STRATEGY 3–6

FOCUS/MOTIVATION

Have students observe once again the chapter-introductory photograph on page O62 and recall the introductory material on page O63. Distribute samples of various fruits and allow students to name the fruits by smell with their eyes closed. Point out that the flavors and odors of many fruits can be manufactured synthetically in the laboratory. These artificial flavor-

ings are called esters. An ester is an organic compound known specifically as a substituted hydrocarbon.

Write the words *methyl salicylate* on the chalkboard. Point out that methyl salicylate is an ester with a very distinctive odor. Challenge students to name the odor (wintergreen).

CONTENT DEVELOPMENT

Point out that although substituted hydrocarbons are organic compounds, they differ from previously studied compounds in that substituted hydrocarbons are formed when one or more hydrogen atoms in a hydrocarbon chain or ring is replaced by a different atom or group of atoms.

Alcohols

Alcohols are substituted hydrocarbons in which one or more hydrogen atoms have been replaced by an –OH group, or hydroxyl group. The simplest alcohol is methanol, CH_3OH. You can see from Figure 3–30 that methanol is formed when 1 hydrogen atom in methane is replaced by the –OH group. Methanol is used to make plastics and synthetic fibers. It is also used in automobile gas tank de-icers to prevent water that has condensed in the tank from freezing. Another important use of methanol is as a solvent. Methanol, however, is very poisonous—even when used externally.

As you can tell from the name methanol, alcohols are named by adding the suffix -ol to the name of the corresponding hydrocarbon. When an –OH group is substituted for 1 hydrogen atom in ethane, the resulting alcohol is ethanol, C_2H_5OH. Ethanol is produced naturally by the action of yeast or bacteria on the sugar stored in grains such as corn, wheat, and barley.

Ethanol is a good solvent for many organic compounds that do not dissolve in water. Ethanol is used in medicines. It is also the alcohol used in alcoholic beverages. In order to make ethanol available for industrial and medicinal uses only, it must be made unfit for beverage purposes. So poisonous compounds such as methanol are added to ethanol. The resulting mixture is called denatured alcohol.

Figure 3–30 *Methanol, an organic alcohol, is used to make plastics, such as the brightly colored insulation on telephone wires.*

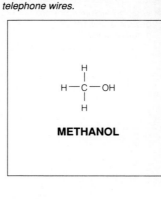

METHANOL

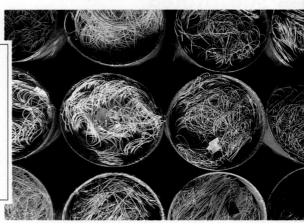

3–6 (continued)

CONTENT DEVELOPMENT

Point out that alcohols are substituted hydrocarbons in which one or more hydrogen atoms have been replaced by an –OH (or hydroxyl) group. Direct students' attention to the structural formula for a methane molecule illustrated in Figure 3–30. It will be easier for students to recognize the change from a hydrocarbon to a substituted hydrocarbon if they had a hydrocarbon molecule to compare the methanol molecule against. Use the chalkboard to draw the structural formula for methane (CH_4) and have students compare each structural formula. Point out that the structural formulas are identical except for an –OH substitution in methanol for a hydrogen atom in methane.

Point out that alcohols such as methanol, ethanol, and phenol are valuable as solvents and also have other important applications in industry and medicine.

ENRICHMENT

Motor-vehicle racing often involves vehicles that are powered by alcohol fuels such as methanol or ethanol. Have interested students use reference materials to discover why racing vehicles use alcohol-based fuels instead of conventional gasolines. Also challenge these students to find out what advantage, if any, alcohol fuels provide a racing vehicle and the reasons why these fuels are not recommended or have not been developed for use in family cars. Have volunteers share their finding with the class.

CONTENT DEVELOPMENT

Point out that organic acids and alcohols are both substituted hydrocarbons. Have students recall that alcohols have one or more hydroxyl groups (–OH) replacing one or more hydrogen atoms. Ex

An alcohol can be in the form of a ring as well as a chain. When 1 hydrogen atom in a benzene ring is replaced by an –OH group, the resulting alcohol is called phenol. Phenol is used in the preparation of plastics and as a disinfectant.

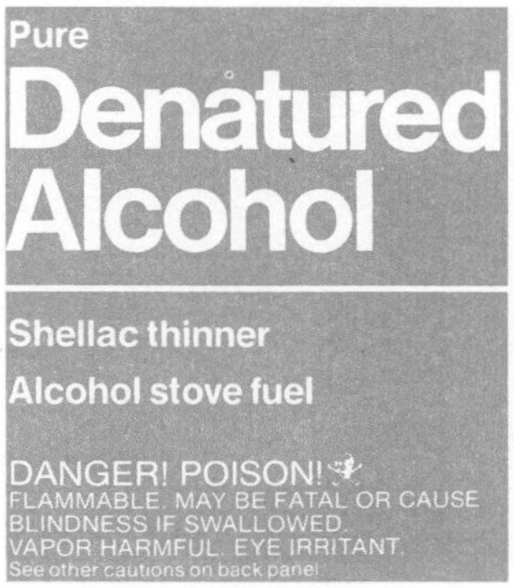

Organic Acids

Organic acids are substituted hydrocarbons that contain the –COOH group, or carboxyl group. Figure 3–32 shows the structural formula for two common organic acids. Notice that one of the carbon-oxygen bonds in the carboxyl group is a double bond.

Organic acids are named by adding the suffix *-oic* to the name of the corresponding hydrocarbon. Most organic acids, however, have common names that are used more frequently. The simplest organic acid is methanoic acid, HCOOH. Methanoic acid is commonly called formic acid. Formic acid is found in nature in the stinging nettle plant and in certain ants. Formic acid produced by an ant causes the ant bite to hurt.

The acid derived from ethane is commonly called acetic acid. Acetic acid is the acid in vinegar. Citric acid, which is found in citrus fruits, is a more complicated organic acid originally derived from the hydrocarbon propane.

Pure
Denatured Alcohol

Shellac thinner

Alcohol stove fuel

DANGER! POISON!
FLAMMABLE. MAY BE FATAL OR CAUSE BLINDNESS IF SWALLOWED. VAPOR HARMFUL. EYE IRRITANT. See other cautions on back panel.

Figure 3–31 *Denatured alcohol is ethanol to which poisonous compounds such as methanol have been added. Why is ethanol denatured?* 1

Figure 3–32 *Formic acid, also known as methanoic acid, is the simplest organic acid. It is the acid produced by ants and is responsible for the pain caused by an ant bite. Acetic acid is the acid in vinegar. What is another name for acetic acid?* 2

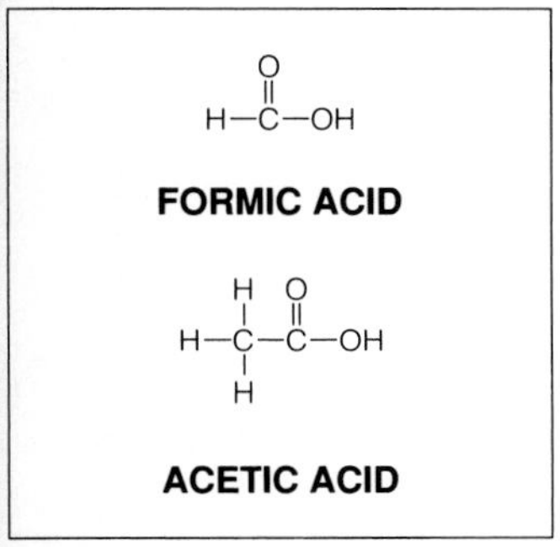

O
‖
H–C–OH

FORMIC ACID

H   O
| ‖
H–C–C–OH
|
H

ACETIC ACID

O ■ 89

Explain that organic acids are different in that they contain the substitution –COOH, or carboxyl group. Have students refer to Figure 3–32, noting the substitution of the –COOH group. Have students compare the molecule of formic acid to the molecule of methane in Figure 3–19 and the molecule of methanol in Figure 3–30, noticing the substitutions as well as the fact that one of the carbon-oxygen bonds in the substituted hydrocarbons is formic acid and acetic acid is a double bond.

● ● ● ● **Integration** ● ● ● ●

Use the relationship for formic acid to ants to integrate zoology concepts into your lesson.

ENRICHMENT

Liquid hydrogen is often used as a rocket fuel. In the future, hydrogen may be used to power vehicles other than rockets. Our transportation industry relies heavily on fossil fuels. The burning of fossil fuels is contributing to global warming and acid rain. Combined with the fact that our fossil fuel supply is finite, hydrogen may become the popular fuel of choice in the future. Hydrogen would offer an abundant and clean-burning fuel.

ECOLOGY NOTE
HARMFUL HYDROCARBONS

Some halogenated hydrocarbons have been proven harmful to our environment. Freon, halogenated methane CCl_2F_2, can decompose and react with ozone (O_3), converting it to molecular oxygen (O_2). The loss of ozone in our upper atmosphere means that more harmful ultraviolet rays may reach the Earth's surface. Ozone screens out many of these harmful rays.

3–6 (continued)

GUIDED PRACTICE

Skills Development

Skill: Making comparisons

• **What do organic acids and organic alcohols have in common?** (Both contain –OH groups.)
• **Write out the structural formula of phenol.** (Check students' drawings.)
• **Write out the structural formula of benzoic acid.** (Check students' drawings. Point out that benzoic acid can be used to make a food preservative.)
• **What is the major difference between an acid and an alcohol?** (An acid contains a carbon double-bonded to an oxygen; an alcohol does not.)
• **Acetic acid is also known as ethanoic**

Figure 3–33 *Halogen derivatives have a variety of uses. Polyvinyl chloride is used to make all sorts of water-repellent gear.*

ACTIVITY
DISCOVERING

Saturated and Unsaturated Fats

Many of the foods you eat contain saturated and unsaturated fats. Using books and other reference materials in the library, find out what these terms mean. Find out the health risks and benefits associated with eating each.

Make a list of foods that contain saturated fats and those that contain unsaturated fats. Analyze your daily diet for a one-week period. Indicate which types of fats you eat, how much of each you eat, and how often.

■ Do you need to modify your daily diet? If so, how?

Esters

If an alcohol and an organic acid are chemically combined, the resulting compound is called an ester. Esters are noted for their pleasant aromas and flavors. The substances you read about in the chapter opener that give flavor to ice cream are esters.

Many esters occur naturally. Fruits such as strawberries, bananas, and pineapples get their sweet smell from esters. Esters can also be produced in the laboratory. Synthetic esters are used as perfume additives and as artificial flavorings.

Halogen Derivatives

Hydrocarbons can undergo substitution reactions in which one or more hydrogen atoms are replaced by an atom or atoms of fluorine, chlorine, bromine, or iodine. The family name for these elements is halogens. So substituted hydrocarbons that contain halogens are called halogen derivatives.

A variety of useful substances result from adding halogens to hydrocarbons. The compound methyl chloride, CH_3Cl, is used as a refrigerant. Tetrachloroethane, $C_2H_2Cl_4$, which consists of 4 chlorine atoms substituted in an ethane molecule, is used in dry cleaning.

When 2 hydrogen atoms in a methane molecule are replaced by chlorine atoms and the other 2 hydrogen atoms are replaced by fluorine atoms, a compound commonly known as Freon, CCl_2F_2, is formed. The actual name of this halogen derivative is dichlorodifluoromethane. Freon is the coolant used in many refrigerators and air conditioners.

3–6 Section Review

1. What is a substituted hydrocarbon?
2. What is an alcohol? An organic acid? An ester?

Critical Thinking—*Identifying Relationships*
3. Methanol is used in car de-icers. What does this tell you about the freezing point of methanol? What role does methanol play in terms of the solution of water and methanol?

acid. Using Figure 3–32, write out the structural formula of ethanoic acid. Then write out the structural formula of ethanol. (Check students' drawings. Have students compare and note the similarities and differences between the drawings.)

CONTENT DEVELOPMENT

Point out that esterification reactions combine an acid and an alcohol chemically. In addition to forming an ester compound, water is released. For example, the odor of oranges can be formed by combining acetic acid and octyl alcohol. Show students the equation involving structural formulas for these molecules so that they can better understand where the water comes from.

● ● ● ● **Integration** ● ● ● ●

Use the introduction of the halogen family of hydrocarbons to integrate halogen concepts into your lesson.

Use the description of Freon to integrate refrigeration concepts into your lesson.

CONNECTIONS

A Chemical You ❸

The human body is one of the most amazing chemical factories ever created. It can produce chemicals from raw materials, start complex chemical reactions, repair and reproduce some of its own parts, and even correct its own mistakes.

What is the fuel that keeps your human chemical factory going? *Nutrients* contained in the foods you eat maintain the proper functioning of all the systems of the body. The three main types of nutrients—carbohydrates, fats and oils, and proteins—are organic compounds.

Carbohydrates are organic molecules of carbon, hydrogen, and oxygen. Carbohydrates are the body's main source of energy. Carbohydrates can be either sugars or starches. The simplest carbohydrate is the sugar glucose, $C_6H_{12}O_6$. The more complex sugar sucrose, $C_{12}H_{22}O_{11}$, is common table sugar.

Starches are another kind of carbohydrate. Starches are made of long chains of sugar molecules hooked together. Starches are found in foods such as bread, cereal, potatoes, pasta, and rice.

Like carbohydrates, fats and oils contain carbon, hydrogen, and oxygen. These molecules are large, complex esters. Fats and oil store twice as much energy as do carbohydrates.

As a class of organic compounds, fats and oils are sometimes called lipids. Fats are solid at room temperature, whereas oils are liquid. Lipids include cooking oils, butter, and the fat in meat. Although fats and oils are high-energy nutrients, too much of these substances can be a health hazard. Unused fats are stored by the body. This increases body weight. In addition, scientific evidence indicates that eating too much animal fat may contribute to heart disease.

Proteins are used to build and repair body parts. Every living part of your body contains proteins. Blood, muscles, brain tissue, skin, and hair all contain proteins. Proteins contain carbon, hydrogen, oxygen, and nitrogen. Some also contain sulfur and phosphorus. Meat, fish, dairy products, and soybeans are sources of proteins.

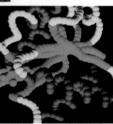

A computer-generated model of a human protein molecule

All foods contain three basic nutrients: carbohydrates, fats and oils, and proteins. A balanced diet should provide the proper amounts of each nutrient.

O ■ 91

ANNOTATION KEY

Integration

❶ Physical Science: Halogens. See *Matter: Building Block of the Universe*, Chapter 5.

❷ Physical Science: Refrigeration. See *Heat Energy*, Chapter 2.

❸ Life Science: Nutrition. See *Human Biology and Health*, Chapter 3.

CONNECTIONS

A CHEMICAL YOU

This feature will help students relate the importance of chemical compounds to their lives. As you discuss this information with students, ask them to describe the ways in which chemistry (especially their studies from the chapter) influences the human body. For example, do they feel that digestive juices would be acidic, basic, or neutral?

If you are teaching thematically, you may want to use the Connections feature to reinforce the themes of energy, patterns of change, scale and structure, systems and interactions, and stability.

Integration: Use the Connections feature to integrate nutrition into your lesson.

INDEPENDENT PRACTICE

Section Review 3–6

1. A substituted hydrocarbon is a hydrocarbon in which one or more hydrogen atoms has been replaced by a different atom or group of atoms. Alcohols are formed by the substitution of one or more hydroxyl groups. Organic acids are formed by the substitution of a carboxyl group. Esters are formed by chemically combining an alcohol and an organic acid. Halogen derivatives are typically formed by the substitution of one or more atoms of fluorine, chlorine, bromine, or iodine.

2. An alcohol is a substituted hydrocarbon in which one or more hydrogen atoms has been replaced by an –OH, or hydroxyl, group; an organic acid is a substituted hydrocarbon in which a hydrogen atom has been replaced by a –COOH, or carboxyl, group; an ester is the chemical combination of an alcohol and an organic acid.

3. The freezing point of methanol is lower than the freezing point of water. The freezing point of a solution of methanol and water is lower than the freezing point of water.

REINFORCEMENT/RETEACHING

Review students' responses to the Section Review questions. Reteach any material that is still unclear, based on students' responses.

CLOSURE

▶ *Review and Reinforcement Guide*

Students may now complete Section 3–6 in the *Review and Reinforcement Guide*.

O ■ 91

Laboratory Investigation

ACIDS, BASES, AND SALTS

BEFORE THE LAB

1. Gather all equipment at least one day prior to the investigation. You should gather enough equipment to meet your class needs, assuming up to six students per group.
2. Prepare dilute solutions of the acids and bases listed. Precision as to concentration is not at all crucial. Simply dilute the concentrated acids in roughly a 15:1 or 20:1 water : acid ratio, adding acid to water. Prepare the bases by weighing out roughly 50 g of each base per liter of solution.
3. Label the bottles.

PRE-LAB DISCUSSION

Have students read the complete laboratory procedure. Before beginning the investigation, avoid discussing with students the general properties of acids, bases, and salts, as this is the topic they are setting out to explore. Lead students, however, to the hypothesis that substances that are classified in the same group, such as acids, should be expected to have certain properties in common. Then discuss the experiment in terms of scientific method, focusing students' attention on the problem, the hypothesis that they have formulated, and the variables they will manipulate in order to test the hypothesis.
• **What safety procedures are to be followed when working with acids and bases?** (Care should be taken when moving acids and bases—spills can burn skin. Any spill on clothing or skin should be immediately washed off with cold water. Also, safety goggles should be worn at all times, and the instructor should be notified of even the smallest spill.)

Laboratory Investigation

Acids, Bases, and Salts

Problem

What are some properties of acids and bases? What happens when acids react with bases?

Materials *(per group)*

safety goggles	test-tube rack
beaker	red and blue
solutions of	litmus paper
H_2SO_4, HCl,	stirring rod
HNO_3	medicine drop-
solutions of	per
KOH, NaOH,	phenolphthalein
$Ca(OH)_2$	evaporating
6 medium-sized	dish
test tubes	paper towels

Procedure 🧪 📦 👁

A. *Acids*

1. Put on your safety goggles. Over a sink, pour about 5 mL of each acid into separate test tubes. **CAUTION:** *Handle acids with extreme care. They can burn the skin.* Place the test tubes in the rack. Test the effect of each acid on litmus paper by dipping a stirring rod into the acid and then touching the rod to the litmus paper. Test each acid with both red and blue litmus paper. *Be sure to clean the rod between uses.* Record your observations.
2. Add 1 drop of phenolphthalein to each test tube. Record your observations.

B. *Bases*

1. Over a sink, pour about 5 mL of each base into separate test tubes. **CAUTION:** *Handle bases with extreme care.* Place the test tubes in the rack. Test the contents of each tube with red and blue litmus paper. Record your observations.
2. Add 1 drop of phenolphthalein to each test tube. Record your observations.

3. Place 5 mL of sodium hydroxide solution in a small beaker and add 2 drops of phenolphthalein. Record the solution's color.
4. While slowly stirring, carefully add a few drops of hydrochloric acid until the mixture changes color. Record the color change. This point is known as the indicator endpoint. Test with red and blue litmus paper. Record your observations.
5. Carefully pour some of the mixture into a porcelain evaporating dish. Let the mixture evaporate until it is dry.

Observations

1. What color do acids turn litmus paper? Phenolphthalein?
2. What color do bases turn litmus paper? Phenolphthalein?
3. What happens to the color of the sodium hydroxide-phenolphthalein solution when hydrochloric acid is added?
4. Does the substance formed by the reaction of sodium hydroxide with hydrochloric acid affect litmus paper?
5. Describe the appearance of the substance that remains after evaporation.

Analysis and Conclusions

1. What are some properties of acids? Of bases?
2. What type of substance is formed when an acid reacts with a base? What is the name of this reaction? What is the other product? Why does this substance have no effect on litmus paper?
3. What is meant by an indicator's endpoint?
4. **On Your Own** Write a balanced equation for the reaction between sodium hydroxide and hydrochloric acid.

SAFETY TIPS

Make sure students wear safety goggles whenever they are working with acids and/or bases. Instruct students on the proper way to handle and dispose of such chemicals. If students should spill chemicals onto their skin or clothing, the chemicals should be immediately washed off with cold water.

TEACHING STRATEGIES

1. It is essential that students be made aware that they must clean each stirring rod between uses.
2. During the investigation, circulate through the work area to ensure that all students are following sensible safety procedures.

Study Guide

Summarizing Key Concepts

3–1 Solution Chemistry
▲ A solution consists of a solute and a solvent.

▲ Solubility is a measure of how much of a particular solute can be dissolved in a given amount of solvent at a certain temperature.

▲ Solutions can be unsaturated, saturated, or supersaturated.

3–2 Acids and Bases
▲ Acids taste sour, turn litmus paper red and phenolphthalein colorless, and ionize in water to form hydrogen ions (H⁺).

▲ Bases feel slippery, taste bitter, turn litmus paper blue and phenolphthalein pink, and produce hydroxide ions (OH⁻) in solution.

3–3 Acids and Bases in Solution: Salts
▲ A neutral substance has a pH of 7. Acids have pH numbers lower than 7. Bases have pH numbers higher than 7.

▲ When an acid chemically combines with a base, the reaction is called neutralization. The products of neutralization are a salt and water.

3–4 Carbon and Its Compounds
▲ Most compounds that contain carbon are called organic compounds.

▲ Isomers have the same molecular formula but different structural formulas.

3–5 Hydrocarbons
▲ The alkanes are saturated hydrocarbons. The alkenes and the alkynes are unsaturated hydrocarbons.

▲ Aromatic hydrocarbons have a ring structure containing 6 carbon atoms.

3–6 Substituted Hydrocarbons
▲ Substituted hydrocarbons include alcohols, organic acids, esters, and halogen derivatives.

Reviewing Key Terms

3–1 Solution Chemistry
solution
solute
solvent
electrolyte
nonelectrolyte
solubility
concentration
concentrated solution
dilute solution
saturated solution
unsaturated solution
supersaturated solution

3–2 Acids and Bases
acid
base

3–3 Acids and Bases in Solution: Salts
pH   neutralization
salt

3–4 Carbon and Its Compounds
organic compound   isomer
structural formula

3–5 Hydrocarbons
hydrocarbon   alkane
saturated hydrocarbon   alkene
unsaturated hydrocarbon   alkyne

3–6 Substituted Hydrocarbons
substituted hydrocarbon

O ■ 93

DISCOVERY STRATEGIES

Discuss how the investigation relates to the chapter by asking open questions similar to the following:

• **Why can't certain acids, bases, and salts be stored in any kind of container?** (The reactive nature of certain acids, bases, and salts will cause some containers to deteriorate, posing a health risk—inferring, applying.)

• **What determines whether a solution will be a relatively good or a relatively poor conductor of electricity?** (Solutions that contain high concentrations of ions are typically good conductors of electricity—applying, relating.)

• **Which of the solutions in this investigation would you predict are electrolytic in nature and which solutions would you predict are nonelectrolytic in nature?** (Accept all predictions—predicting, comparing.)

Chapter Review

ALTERNATIVE ASSESSMENT

The *Prentice Hall Science* program includes a variety of testing components and methodologies. Aside from the Chapter Review questions, you may opt to use the Chapter Test or the Computer Test Bank Test in your *Test Book* for assessment of important facts and concepts. In addition, Performance-Based Tests are included in your *Test Book*. These Performance-Based Tests are designed to test science process skills, rather than factual content recall. Since they are not content dependent, Performance-Based Tests can be distributed after students complete a chapter or after they complete the entire textbook.

CONTENT REVIEW

Multiple Choice

1. c
2. b
3. a
4. a
5. a
6. b
7. c
8. b
9. c
10. d

True or False

1. F, solute
2. T
3. F, good
4. F, 7
5. T
6. F, saturated
7. T
8. T
9. F, isomers

Concept Mapping

Row 1: Electrolyte; Freezing point depression
Row 2: Solutions
Row 3: Nonelectrolyte
Row 4: Solute; may be; Unsaturated
Row 5: Supersaturated

CONCEPT MASTERY

1. The rate of solution can be increased for a solid in a liquid by stirring, by heating, and by powdering; stirring vs. not stirring, in hot tea vs. cold tea, a sugar cube vs. powdered sugar.

Content Review

Multiple Choice

Choose the letter of the answer that best completes each statement.

1. A solution that conducts an electric current is called a(an)
   a. nonelectrolyte. c. electrolyte.
   b. tincture.   d. colloid.
2. Which process will not increase the rate of solution of a solid in a liquid?
   a. powdering the solution
   b. cooling the solution
   c. heating the solution
   d. stirring the solution
3. A solution that contains all the solute it can hold at a given temperature is said to be
   a. saturated.   c. supersaturated.
   b. unsaturated.   d. dissociated.
4. Which ion do bases contain?
   a. OH^- b. H_3O^+ c. H^+ d. NH_4^+
5. Organic compounds always contain
   a. carbon.   c. halogens.
   b. oxygen.   d. carboxyl groups.

6. The pH of the products formed by a neutralization reaction is
   a. 1.   b. 7.   c. 14.   d. 0.
7. The type of bonding found in organic compounds is
   a. metallic.   c. covalent.
   b. ionic.   d. coordinate.
8. A compound that contains only carbon and hydrogen is called a(an)
   a. isomer.   c. carbohydrate.
   b. hydrocarbon.   d. alcohol.
9. The simplest aromatic hydrocarbon is
   a. cyclohexane.   c. benzene.
   b. methane.   d. phenol.
10. The –OH group is characteristic of an
    a. organic acid.
    b. aromatic compound.
    c. ester.
    d. alcohol.

True or False

If the statement is true, write "true." If it is false, change the underlined word or words to make the statement true.

1. In a solution, the substance being dissolved is the <u>solvent</u>.
2. Acids are often defined as <u>proton donors</u>.
3. Strong acids are <u>poor</u> electrolytes.
4. A neutral solution has a pH of <u>10</u>.
5. In <u>neutralization</u>, an acid reacts with a base.
6. Hydrocarbons that contain only single bonds are said to be <u>unsaturated</u>.
7. The 4-carbon alkane is called <u>butane</u>.
8. An organic acid is characterized by the group <u>–COOH</u>.
9. Compounds that have the same molecular formula but different structural formulas are called <u>isotopes</u>.

Concept Mapping

Complete the following concept map for Section 3–1. Refer to pages O6–O7 to construct a concept map for the entire chapter.

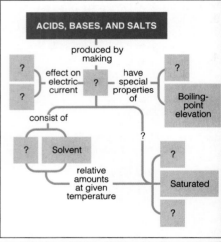

---

2. Saturated means holding all the solute it can hold at a given temperature. If the amount is small compared to the amount of solvent, however, the solution would be saturated but dilute. If the amount necessary to make it saturated is great compared to the amount of solvent, then the solution is concentrated.

3. Because water is a polar covalent compound, it can serve as both a proton donor and a proton acceptor. Thus it can sometimes be an acid and sometimes be a base.

4. Carbon's ability to combine with itself in single, double, or triple covalent bonds; carbon's ability to combine with other elements; the fact that carbon compounds form isomers; isomers can form straight chains, branched chains, or ring structures.

5. A structural formula shows the kind, number, and arrangement of atoms in a molecule.

6. Alcohols do not ionize in water, producing an –OH ion. Therefore, they are not proton acceptors, or bases.

Concept Mastery

Discuss each of the following in a brief paragraph.

1. Describe three ways in which the rate of solution of a solid in a liquid can be increased. Use a specific example.
2. Explain how a saturated solution of one solute can be concentrated and a saturated solution of a different solute can be dilute.
3. Explain why water is both an acid and a base.
4. Discuss four reasons why carbon compounds are so abundant.
5. Explain the importance of structural formulas in organic chemistry.
6. In this chapter you learned that compounds called bases contain the hydroxide ion, $-OH^-$. Alcohols also contain the $-OH$ group. Why are alcohols not bases?

Critical Thinking and Problem Solving

Use the skills you have developed in this chapter to answer each of the following

1. **Classifying compounds** Identify each of the following compounds as an acid, base, or salt:
   a. $CaCO_3$ c. $CsOH$ e. $MgSO_4$
   b. HI d. H_3PO_4 f. $Ga(OH)_3$
2. **Applying concepts** The caps are removed from a warm bottle and a cold bottle of carbonated beverage. The soda in the cold bottle fizzes slightly. The soda in the warm bottle fizzes rapidly.
   a. What two conditions affecting solubility are present here?
   b. Which condition is the variable?
   c. Give an explanation for what happens.
3. **Designing an experiment** Describe an experiment to determine whether a solution is saturated or unsaturated.
4. **Identifying patterns** A crystal of solute is added to a saturated, unsaturated, and supersaturated solution. Describe what happens in each case.
5. **Classifying hydrocarbons** Classify each of the following hydrocarbons as an alkane, alkene, alkyne, or aromatic compound.
   a. C_4H_{10} c. C_2H_4 e. $C_{13}H_{26}$
   b. C_3H_4 d. C_6H_6 f. $C_{42}H_{82}$
6. **Applying definitions** Draw the structural formulas for the isomers of hexane.
7. **Making diagrams** Draw structural formulas for the following compounds:
   a. hexane
   b. butene
   c. propyne
8. **Classifying substituted hydrocarbons** Classify each of the following compounds as a(an) ester, alcohol, organic acid, or halogen derivative:
   a. $C_2H_5COOC_3H_7$ c. C_6H_5OH
   b. C_4H_9Cl d. C_2H_5COOH
9. **Identifying patterns** Using the general formulas for the alkanes, alkenes, and alkynes, show why the number of hydrogen atoms in each series decreases by two.
10. **Drawing a conclusion** Explain why the alkene series and the alkyne series begin with a 2-carbon hydrocarbon rather than a 1-carbon hydrocarbon, as the alkane series does. Use structural formulas to support your explanation.
11. **Using the writing process** Water is often called the universal solvent. Is it truly? Write a short story about whether a true universal solvent exists. Describe all its physical properties. End your story by describing how you would bring a sample of it to class.

---

each series decreases by two because of a change in bonding from a single to double to triple. The addition of another carbon-to-carbon bond eliminates 2 hydrogen atoms.

10. The alkene has a double bond between 2 adjacent carbon atoms, so the minimum number of carbon atoms has to be 2. The alkyne has a triple bond, so the minimum number of carbon atoms has to be 2. Check students' drawings.

11. If such a solvent existed, there would be no container to hold it because the solvent would dissolve any container.

KEEPING A PORTFOLIO

You might want to assign some of the Concept Mastery and Critical Thinking and Problem Solving questions as homework and have students include their responses to unassigned questions in their portfolio. Students should be encouraged to include both the questions and the answer in their portfolio.

ISSUES IN SCIENCE

The following issues can be used as springboards for discussion or given as writing assignments:

1. Acidic substances released into the atmosphere by certain industries can eventually result in acid rain, which is destructive to many plant and animal life forms. What can be done about this problem? (Answers will vary. Some students may suggest adding basic substances to bodies of water acidified by such rain, in order to neutralize the acids. Others may suggest more efficient "cleansing" of such airborne industrial wastes or the development of other ways of disposing of them.)

2. Many people believe that life based on hydrocarbons may exist on other planets. Others disagree. Still others believe that silicon-based organisms may exist on other worlds. What is your opinion?

CRITICAL THINKING AND PROBLEM SOLVING

1. Acid: b, d; base: c, f; salt: a, e.
2a. Temperature and pressure; b. temperature; c. As temperature increases, the solubility of carbon dioxide in the beverage decreases. The beverage in the warm bottle effervesces rapidly because the solubility of the carbon dioxide gas is decreased.
3. Students' experiments should be logical, including a control and a variable. Students should infer that by adding more solute they can determine if the solution is saturated.
4. Saturated: solute falls to bottom; Unsaturated: dissolves; Supersaturated: crystal will cause all excess solute to precipitate out.
5. Alkane: a; alkene: c, e; alkyne: b, f; aromatic: d.
6. Check students' drawings.
7. Check student's drawings.
8. Ester: a; alcohol: c; organic acid: d; halogen derivative: b.
9. The number of hydrogen atoms in

Chapter 4   PETROCHEMICAL TECHNOLOGY

SECTION	HANDS-ON ACTIVITIES
4–1 What Is Petroleum? pages O98–O101 Multicultural Opportunity 4–1, p. O98 ESL Strategy 4–1, p. O98	**Student Edition** ACTIVITY (Doing): Where Is the Oil? p. O98 ACTIVITY BANK: Oil Spill, p. O157 **Laboratory Manual** Fractional Distillation, p. O57 **Activity Book** CHAPTER DISCOVERY: Making Aspirin, p. O87 ACTIVITY: A Model Oil Reservoir, p. O95 **Teacher Edition** Petrochemical Products, p. O96d
4–2 Petrochemical Products pages O101–O105 Multicultural Opportunity 4–2, p. O101 ESL Strategy 4–2, p. O101	**Student Edition** ACTIVITY (Doing): Homemade Adhesives, p. O103 LABORATORY INVESTIGATION: Comparing Natural and Synthetic Polymers, p. O106 **Laboratory Manual** Tensile Strength of Natural and Synthetic Polymers, p. O53 **Product Testing Activity** Testing Nail Enamels **Teacher Edition** Making Polymer Nylon, p. O96d Testing Synthetics, p. O96d
Chapter Review pages O106–O109	

OUTSIDE TEACHER RESOURCES

Books

Cantow, H. J. (ed). *Polymer Chemistry,*
  Springer-Verlag.
Driver, W. E. *Plastics Chemistry and
  Technology,* Van Nostrand Reinhold.
Jenkins, A. D. *Polymer Science,* Elsevier.

Audiovisuals

Man-Made Macromolecules: Polymers, video,
  The Media Guild
Mechanism of an Organic Reaction, film,
  CHEM Study
Synthesis of an Organic Compound, film,
  CHEM Study

OTHER ACTIVITIES	MEDIA AND TECHNOLOGY
Student Edition ACTIVITY (Reading): All About Oil, p. O101 **Activity Book** ACTIVITY: Fractional Distillation of Petroleum, p. O97 **Review and Reinforcement Guide** Section 4–1, p. O39	**English/Spanish Audiotapes** Section 4–1
Activity Book ACTIVITY: A World of Synthetic Polymers, p. O89 ACTIVITY: Constructing a Nylon Polymer Molecule, p. O91 ACTIVITY: Bakelite, p. O93 **Review and Reinforcement Guide** Section 4–2, p. O43	**Interactive Videodisc** ScienceVision: Chemical Pursuits **English/Spanish Audiotapes** Section 4–2
Test Book Chapter Test, p. O77 Performance-Based Tests, p. O117	**Test Book** Computer Test Bank Test, p. O83

*All materials in the Chapter Planning Guide Grid are available as part of the Prentice Hall Science Learning System.

CHAPTER OVERVIEW

Many products come from petroleum. These products include fuels, such as heating oil and gasoline, and synthetic substances, such as nylon, plastics, and synthetic rubber.

Raw petroleum, which is called crude oil, contains many different parts, or fractions. These fractions can be separated by a process known as fractional distillation.

Many products that we use every day are made from petrochemicals. Petrochemicals are produced by polymerization, a process by which many molecules are chemically bonded together to form long chains called polymers.

Although some polymers occur in nature, many more are synthetic. Natural polymers include silk, cotton, and cellulose. Synthetic polymers are designed to be flexible, strong, lightweight, and long lasting. They include rayon, plastics, and synthetic rubber.

4–1 WHAT IS PETROLEUM?
THEMATIC FOCUS

The purpose of this section is to introduce students to the process used to separate petroleum into its useful parts. This process is known as fractional distillation. Students will learn that fractional distillation takes place in a fractionating tower, where crude oil is headed to a temperature of about 385°C.

Students will learn about the various products that are produced from a barrel of crude oil. They will come to understand that because each fraction of petroleum has a different boiling point, raw petroleum can be vaporized and then condensed to draw off each separate substance.

The themes that can be focused on in this section are evolution, energy, systems and interactions, and stability.

Evolution: Petroleum is believed to have formed from the remains of dead animals and plants buried in the oceans and subjected to tremendous heat and pressure over millions of years.

*Energy: Petroleum is particularly important because a tremendous amount of energy is released when petroleum is burned.

*Systems and interactions: Petroleum can be divided into its various parts, or fractions, in a process known as fractional distillation.

*Stability: Petroleum is a nonrenewable resource. Its supply is limited, and once it is diminished, no more will be produced.

PERFORMANCE OBJECTIVES 4–1

1. Explain how petroleum is separated into its useful parts by fractional distillation.
2. Describe how each fraction of petroleum separates in a fractionating tower.
3. Describe some useful products made from petroleum.

SCIENCE TERMS 4–1

petroleum p. O98
fraction p. O99
refining p. O99

4–2 PETROCHEMICAL PRODUCTS
THEMATIC FOCUS

The purpose of this section is to introduce students to the many useful products that are made from petrochemicals. These products include plastics, synthetic fibers, and medicines.

Students will learn that petrochemical products are made possible by a process called polymerization. Polymerization involves the chemical bonding of molecules to make gigantic chains called polymers.

Although most of the polymers discussed in this section are synthetic materials made from petroleum, students will also read about natural polymers. These include cotton, wool, cellulose, and proteins.

The themes that can be focused on in this section are patterns of change, scale and structure, and unity and diversity.

*Patterns of change: Raw materials from petroleum are converted into chemicals. The chemicals are then used to make various products called petrochemical products.

*Scale and structure: Polymers are molecules made of chains of smaller molecules called monomers.

Unity and diversity: Some polymers are found naturally; others are produced synthetically from petrochemicals.

PERFORMANCE OBJECTIVES 4–2

1. Define what a polymer is and describe the process of polymerization.
2. Describe some products made from petrochemicals.

SCIENCE TERMS 4–2

petrochemical product p. O101
monomer p. O101
polymer p. O101
natural polymer p. O102
polymerization p. O102
synthetic polymer p. O103

Discovery *Learning*

TEACHER DEMONSTRATIONS
MODELING

Petrochemical Products

Bring in samples of several materials such as plastic food wrap, a piece of Teflon cookware, a pair of pantyhose, a plastic food-storage container, a shirt or blouse made of a synthetic fabric, an aspirin, a rubber washer or other plumbing component, and a container of antifreeze. Make a display of the items and have students observe them.

In what way are these items similar? (Answers will vary, but students should be guided to recognize that all are made from synthetic materials. By the end of the chapter, they will understand that all are made from petrochemicals.)

Encourage students to recognize the diversity of the items by asking:

Can you classify these items according to how each is used? (Food preparation and storage: plastic wrap, plastic container, Teflon cookware; articles of clothing: shirt or blouse, pantyhose; car maintenance: antifreeze; home maintenance: washer; medicine: aspirin.)

Making Polymer Nylon

This demonstration shows the preparation of the synthetic fiber nylon. You will need the following materials: adipyl chloride solution (0.25 M in hexane), hexamethylene diamine solution (0.50 M in 0.50 M NaOH), acetone (in alcohol), paper clip, small beakers, graduated cylinder.

. Bend apart the paper clip, leaving a hook at one end.

. Pour 5 mL hexamethylene diamine solution into one beaker.

. Slowly add 5 mL adipyl chloride solution. A white film should form as the two solutions meet.

. Put the paper clip into the beaker, hooking around the white film. Then slowly lift the paper clip from the beaker.

. This white substance is the polymer nylon. Continue removing the nylon from the beaker until none remains.

. Place the nylon in a beaker and wash it. Rinse with acetone and let it dry.

Testing Synthetics

Display a large piece of thick clear-plastic sheeting (Plexiglas) for students to observe. Do not tell the students it is plastic.

• **Can you hit this with a hammer?** (Most students will say it depends on what kind of material it is.)

• **What would happen if it were glass and you hit it with a hammer?** (The glass would break.)

• **What would happen if it were plastic and you hit it with a hammer?** (Plastic probably would take the hit and would not break.)

Have several students hit the plastic sheet with a hammer.

• **When would this type of material be better than glass?** (Responses might include anytime something might get hit, and when you do not want the glass to break.)

CHAPTER 4
Petrochemical Technology

INTEGRATING SCIENCE

This physical science chapter provides you with numerous opportunities to integrate other areas of science, as well as other disciplines, into your curriculum. Blue numbered annotations on the student page and integration notes on the teacher wraparound pages alert you to areas of possible integration.

In this chapter you can integrate language arts (pp. 98, 101), earth science and geology (p. 98), social studies (p. 98), physical science and lubricants (p. 100), earth science and energy resources (p. 101), life science and physiology (p. 102), physical science and atomic bonding (p. 102), and life science and the excretory system (p. 105).

SCIENCE, TECHNOLOGY, AND SOCIETY/COOPERATIVE LEARNING

Can you go through a day without using something made of plastic? Probably not! Chemical technology has produced many different types of plastic with a variety of uses and properties. Plastic can be tough, rigid, soft, flexible, and can be made into any shape or color. The convenience of plastic has led to an increase in its use since its development, but plastic use has created some problems.

Plastics are complicating the garbage-disposal problem. Because plastic is relatively inexpensive and very flexible, it became the material of choice for our disposable society—from dinnerware to razors to baking dishes. The increased amount of waste material produced by the disposal of items after just one use threatens to consume what little landfill space is available. Disposing of plastics by burning instead of burying creates a potential health hazard: toxic fumes.

Plastics, because they are not biodegradable, pose a hazard to many forms of wildlife. Animals often confuse plastic containers for food. When they eat the plastic, it blocks their digestive system, and they die. Other animals get entangled in plastic trash and are severely hurt or even killed.

INTRODUCING CHAPTER 4

DISCOVERY LEARNING

▶ *Activity Book*

Begin your introduction to this chapter by using the Chapter 4 Discovery Activity from the *Activity Book*. Using this activity, students will discover a relationship between the hydrocarbon ethylene and aspirin.

USING THE TEXTBOOK

Have students examine the photograph on page O96.
• **What is happening in this photograph** (An oil well is on fire.)
• **Why are fires a common problem when drilling for oil?** (Accept all logical answers. Petroleum products are used for fuel and are easily ignited.)

Have students read the chapter-opening text.

Petrochemical Technology

A telephone call at any time of the day or night may summon you to travel to a distant location. Once you arrive at your destination, your life will be in danger almost constantly. The situation will be explosive. One wrong move—even one unfortunate act of nature—could result in a tremendous and fatal explosion. Your role there may last only a day or perhaps several months. What is this awesome-sounding job? It is putting out oil-well fires. Does it sound like a job you might enjoy? Probably not. But one brave and unusual man who goes by the name of Red Adair loves it!

Oil wells, and the liquids and gases within them, are highly flammable. Once an oil fire begins, it will burn until all the fuel is gone. Putting out an oil-well fire requires enormous ingenuity and courage. Nonetheless, Adair has never failed to put out a fire. Some have taken six months and others only thirty seconds.

What is so important about oil that people are willing to risk their lives to find it, extract it from the Earth, and protect it? In this chapter you will learn the answer to that question. You will also find out what oil is and how it can be used.

Journal *Activity*

You and Your World Look around your home at the objects that surround you every day. Do you see a lot of plastic? You can probably find plastic in your closet, on your shelves, in your refrigerator—anywhere you look. In our journal, describe some of the objects you find that are made of plastic. Explain why the plastic is useful and what material may have been used before plastic was developed.

Only the know-how of Red Adair (inset) is enough to match an oil field burning out of control.

O ■ 97

Many of the products used before the development of plastic (paper, aluminum, glass) can be recycled. Most of the plastic replacing these products is not. The change to plastic packaging has increased the amount of raw materials used in production. Some advances in recycling plastics have been made. Polyethylene terephthatlate (PET) bottles—the kind used for soft-drink containers—can be recycled. This recycled plastic is then used for floor coverings, car parts, and pillow stuffing.

Cooperative learning: Using preassigned lab groups or randomly selected teams, have groups complete the following assignment:

• Have groups brainstorm a list of all the plastic products they use in the course of a school day. Have groups place their list in a chart format that can be used to identify alternative materials that could be used in place of the plastic. Groups should identify possible consequences of changing from plastic to another form of material and decide if the consequences are acceptable. For example, disposable plastic silverware could be replaced with metal utensils in the school cafeteria, but this would require additional dishwashing machines and perhaps personnel.

See Cooperative Learning in the *Teacher's Desk Reference.*

JOURNAL ACTIVITY

You may want to use the Journal Activity as the basis of a class discussion. There is such a wide variety of plastic items in the students' environment that you may wish to have them limit their journal descriptions to a maximum of five plastic objects. Students should be instructed to keep their Journal Activity in their portfolio.

What skills and abilities would someone need to be a firefighter? (Accept all logical answers. Students may discuss firefighters who help to protect their homes and other buildings in their community as well as oil-well firefighters such as Red Adair.)

What type of clothing is needed to protect people who put out fires? (Heat-resistant materials.)

Point out that many of the fabrics used in protective clothing are made from petroleum-based chemicals. Red Adair and his team may actually be wearing clothes made from oil!

• **What other products are made from the petroleum extracted from oil wells?** (Accept all logical answers. Explain that the chapter will give many examples of petroleum-based products.)

• **What makes oil such a valuable resource?** (Answers may vary, but students should be guided to recognize that an enormous variety of products can be produced from oil and that many of these products are essential or extremely valuable to our lives.)

Explain that oil is the world's leading energy resource and that oil is also the raw material from which many important synthetic substances are made.

4-1 What Is Petroleum?

Guide for Reading

Focus on these questions as you read.
▸ *What is petroleum?*
▸ *How does fractional distillation separate petroleum into its various components?*

ACTIVITY
DOING

Where Is the Oil?

① Using books and other reference materials in the library, find out where the major oil fields in the world are located. Make a map showing these oil fields. Be sure to include oil fields that have recently been discovered, as well as areas that are currently believed to contain oil.

Figure 4–1 *The sight of thick black liquid shooting high into the air has been a welcomed one since the first oil well was built in Pennsylvania.*

4-1 What Is Petroleum?

Have you ever seen a movie in which a group of people gathered around a well were shouting and cheering as a thick black liquid was rising up through the ground and spurting high into the air? Or have you ever read about geologists who study the composition of the Earth in an attempt to find oil? If so, you may have realized—correctly so—that crude oil is a rather valuable substance. This sought after crude oil, one form of **petroleum,** has been called black gold because of its tremendous importance. Fuels made from petroleum provide nearly half the energy used in the world. And thousands of products—from the bathing suit you wear when swimming to the toothpaste you use when brushing your teeth—are made from this petroleum.

Petroleum is a substance believed to have been formed hundreds of millions of years ago when layers of dead plants and animals were buried beneath sediments such as mud, sand, silt, or clay at the bottom of the oceans. Over millions of years, heat and great pressure changed the plant and animal remains into petroleum. Petroleum is a nonrenewable resource. A nonrenewable resource is one that

TEACHING STRATEGY 4-1

FOCUS/MOTIVATION

Draw on the chalkboard two beakers filled with liquid. Label one beaker Water: boiling point = 100°C. Label the other beaker Carbon Tetrachloride: boiling point = 76.5°C.
• **Suppose you had a mixture of these two liquids and you wanted to separate them. How would you go about it?** (Allow students to suggest whatever plans they wish.)

If no one has suggested taking advantage of the difference in boiling point, ask:
• **What physical property of the two liquids is listed on the chalkboard?** (Boiling point.)
• **How might this property be useful in separating the two liquids?** (One would boil before the other, leaving the other liquid behind.)

Note that although this is true, simply boiling the mixture does not solve the problem, because one liquid will "disappear" as it vaporizes into the air.

CONTENT DEVELOPMENT

Use the Focus/Motivation activity to lead into a discussion of fractional distillation. Write the boiling points for the various elements of petroleum (from the Facts and Figures chart) on the chalkboard.
• **If these are some of the essential components of petroleum, how might they be separated from one another?** (By boiling, or vaporizing, the petroleum.)

annot be replaced once it is used up. There is only certain amount of petroleum in existence. Once he existing petroleum is used up, no more will be vailable.

Despite the huge variety of products obtained rom petroleum, few people ever see the substance tself. The liquid form that gushes from deep within he Earth is a mixture of chemicals called crude oil. Petroleum can also be found as a solid in certain ocks and sand. It has been called black gold because it is usually black or dark brown. But it can e green, red, yellow, or even colorless. Petroleum nay flow as easily as water, or it may ooze slowly— ike thick tar. The color and thickness of petroleum lepend on the substances that make it up.

Separating Petroleum Into Parts

By itself, petroleum is almost useless. But the different parts, or **fractions,** of petroleum are among he most useful chemicals in the world. **Petroleum is eparated into its useful parts by a process called ractional distillation.** The process of distillation nvolves heating a liquid until it vaporizes (changes nto a gas) and then allowing the vapor to cool until t condenses (turns back into a liquid). The different fractions of petroleum have different boiling oints. So each fraction vaporizes at a different emperature than do the others. The temperature t which a substance boils is the same as the temerature at which it condenses. So if each fraction aporizes at a different temperature, then each raction will condense back to a liquid at a different emperature. By removing, or drawing off, each raction as it condenses, petroleum can easily be eparated into its various parts.

Fractional distillation of petroleum is done in a ractionating tower. The process of separating petroeum into its fractions is called **refining.** Refining etroleum is done at a large plant called a refinery. t a refinery, fractionating towers may rise 30 meters r more. Figure 4–3 on page 100 shows a fractionatng tower. Petroleum is piped into the base of the ractionating tower and heated to about 385°C. At this emperature, which is higher than the boiling points f most of the fractions, the petroleum vaporizes.

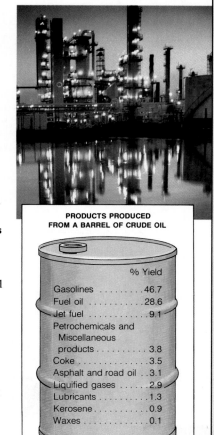

PRODUCTS PRODUCED FROM A BARREL OF CRUDE OIL

	% Yield
Gasolines	46.7
Fuel oil	28.6
Jet fuel	9.1
Petrochemicals and Miscellaneous products	3.8
Coke	3.5
Asphalt and road oil	3.1
Liquified gases	2.9
Lubricants	1.3
Kerosene	0.9
Waxes	0.1

Figure 4–2 *Petroleum is separated into fractions in fractionating columns in an oil refinery. The numbers inside the barrel show the amount, or percentage yield, of each fraction that can be obtained from a barrel of crude oil. Which fraction represents the highest percentage yield?* 1

O ■ 99

ACTIVITY
DOING
WHERE IS THE OIL?

Skills: Applying concepts, making diagrams, making comparisons

Before students begin their research for this activity, have them guess where most of the oil fields in the world are located. Then they can check their speculations as they work on their reports.

Students' maps will include many sources of oil in the Middle East. However, crude oil is also found in North America, the former Soviet Union, and China, as well as in other locations around the world.

Integration: Use this Activity to integrate social studies into your science lesson.

● **What must be done once the mixture has een vaporized?** (The temperature must e lowered in order to condense the vaors into liquids.)

● ● ● ● **Integration** ● ● ● ●

Use the formation of petroleum to integrate geology into your lesson.

Use the illustration of the first oil well n Pennsylvania to integrate social studies concepts into your lesson.

GUIDED PRACTICE

▶ *Laboratory Manual*

Skills Development

Skills: Applying concepts, making observations, making comparisons

At this point you may want to have students complete the Chapter 4 Laboratory Investigation in the *Laboratory Manual* called Fractional Distillation. In the investigation students will perform a fractional distillation of a mixture of water, isopropyl alcohol, and ethylene glycol.

ENRICHMENT

▶ *Activity Book*

Students who have mastered the concepts in this section will be challenged by the Chapter 4 activity called Fractional Distillation of Petroleum.

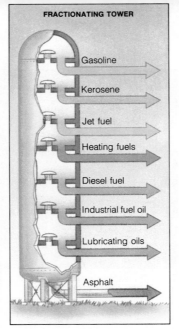

FRACTIONATING TOWER

Gasoline
Kerosene
Jet fuel
Heating fuels
Diesel fuel
Industrial fuel oil
Lubricating oils
Asphalt

100 ■ O

Figure 4–3 *Each fraction of petroleum condenses at a different temperature and is drawn off in collecting vessels located at fixed points along the column. Where in the tower is the temperature lowest? Where is it highest?* ❶

When the petroleum vaporizes, the fractions rise up the tower. As they rise, they cool and condense. Some fractions condense at high temperatures. These fractions condense right away near the bottom of the tower and are drawn off into collecting vessels. Other fractions continue to rise in the tower. These fractions are drawn off at higher levels in the tower. As a result of this vaporization-condensation process, the various fractions of petroleum are separated and collected.

You will notice in Figure 4–3 that asphalt is collected at the bottom of the fractionating tower. Asphalt requires a temperature even higher than 385°C to vaporize. When the other fractions vaporize, asphalt is left behind as a liquid that runs out of the bottom of the tower. Which fraction in the tower condenses at the lowest temperature? ❷

Petroleum Products

Asphalt—the main material used for building roads—is one product that comes directly from petroleum. Wax, used in furniture polish and milk cartons, is another. Asphalt and wax fall into the category of raw materials that come from the separation of petroleum and are used in manufacturing. Many of the other raw materials in this category, however, are converted to chemicals from which a variety of products—ranging from cosmetics to fertilizers—are made. You will read about these products in the next section.

Another group of petroleum products includes lubricants. Lubricants are substances that reduce friction between moving parts of equipment. The oil applied to the gears of a bicycle is an example of a lubricant. Lubricants are used in many machines—from delicate scientific equipment to the landing gear of an aircraft. Can you think of some other uses of lubricants? ❸

Figure 4–4 *The asphalt used to pave roadways comes directly from petroleum. Lubricants used to make machinery run more efficiently also come from petroleum.*

4–1 (continued)

CONTENT DEVELOPMENT

Have students observe the fractionating tower in Figure 4–3.

• **What do you notice about the order of substances in the tower?** (Those with the lowest boiling points are at the top, and those with the highest boiling points are at the bottom.)

• **Why is this the case?** (Those that condense at a lower temperature have more

time to rise in the tower as the vapor is cooled.)

INDEPENDENT PRACTICE

▶ *Activity Book*

Students who need practice with the concepts of this section should be provided with the Chapter 4 activity called A Model Oil Reservoir.

CONTENT DEVELOPMENT

Explain that an important group of petroleum products includes lubricants.

• **Are lubricants used in automobile engines? Why?** (Yes. Because lubrication is needed to reduce friction between the moving parts of an engine.)

• **What would happen if you did not keep an automobile engine properly lubricated?** (The engine would freeze up because friction between the moving parts would create extreme heat. This heat would make the parts expand and would create more friction.)

The greatest percentage of petroleum products includes fuels. Fuels made from petroleum burn easily and release a tremendous amount of energy, primarily in the form of heat. They are also easier to handle, store, and transport than are other fuels, such as coal and wood. Petroleum is the source of nearly all the fuels used for transportation and the many fuels used to produce heat and electricity.

4–1 Section Review

1. What is petroleum and why is it important?
2. Describe the process of fractional distillation.
3. Describe the products produced from petroleum.

Critical Thinking—*Applying Concepts*

4. How would you separate three substances—A, B, and C—whose boiling points are 50°C, 100°C, and 150°C, respectively?

4–2 Petrochemical Products

Paint a picture, pour milk from a plastic container, or put on a pair of sneakers and you are using a product made from petroleum, or a **petrochemical product.**

Polymer Chemistry

The petrochemical products that are part of your life come from the chemicals produced from petroleum. (The word petrochemical refers to chemicals that come from petroleum.) Petrochemical products usually consist of molecules that take the form of long chains. Each link in the chain is a small molecular unit called a **monomer.** (The prefix *mono-* means singular, or one.) The entire molecule chain is called a **polymer.** (The prefix *poly-* means many.)

Figure 4–5 A polymer is made up of a series of monomers. What factors distinguish one polymer from another?

ACTIVITY READING

All About Oil

Some discoveries in history are acts of genius. Others are accidents. But most are a little of both. Read Isaac Asimov's *How Did We Find Out About Oil?* to discover how oil was first discovered, extracted, and used.

Activity Bank

Oil Spill, p.157

Guide for Reading

Focus on these questions as you read.
▶ *What is a petrochemical product?*
▶ *What are some products of polymerization?*

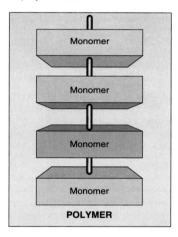

Monomer

Monomer

Monomer

Monomer

POLYMER

4–2 Petrochemical Products

sation temperatures to separate substances, could be used.

TEACHING STRATEGY 4–2

BACKGROUND INFORMATION
POLYMERIZATION REACTIONS

There are two types of polymerization reactions. The first of these is addition polymerization. In addition polymerization, bonding between monomers is accomplished by opening the double bonds between carbon atoms in molecules such as ethylene. Polymers such as Teflon and plexiglass are made by this process.

The second type of polymerization reaction is called condensation polymerization. In this type of reaction, atoms must be eliminated in order for the bonds to form. The polymer Dacron is formed by the condensation process.

HISTORICAL NOTE
SYNTHETIC RUBBER

Synthetic rubber was invented in the United States during World War II. Rubber was an important substance in the making of military materials. Our supply of natural rubber was cut off because of the war. The first synthetic rubber tires were acceptable but not as good as natural rubber. Today, due to continued research and technological development, synthetic rubber products are far superior to natural rubber products.

Figure 4–6 *Silk is spun from threads of silk worm cocoons. Silk is a natural polymer. Rubber obtained from rubber trees is also a natural polymer.*

Figure 4–7 *Synthetic fibers, such as nylon, are used to make the carpets you walk on. What are some other uses of synthetic fibers?* ❶

102 ■ O

The types of monomers and the length and shape of the polymer chain determine the physical properties of the polymer. Manufacturers of petrochemical products join monomers together to build polymers. A general term for this process is polymer chemistry.

Natural Polymers

Most of the polymers you will read about in this chapter are made from petrochemicals. Some polymers, however, do occur in nature. Cotton, silk, wool, and natural rubber are all **natural polymers.** Cellulose and lignin, which are important parts of wood, are natural polymers. In fact, all living things contain polymers. Yes—that includes you! Protein, an essential ingredient of living matter, is a polymer. The monomers from which proteins are made are called amino acids. Combined in groups of one hundred or more units, amino-acid monomers form many of the parts of your body—from hair to heart muscle.

Synthetic Polymers

The first polymer was manufactured in 1909. Since then, **polymerization** (poh-lihm-er-uh-ZAY-shuhn) has come a long way. Polymerization is the process of chemically bonding monomers to form polymers. Most early polymers consisted of fewer than two hundred monomers. Today's polymers may contain thousands of monomers. The many ways in which these monomers can be linked may be very complex. They include single chains, parallel chains, intertwining chains, spirals, loops, and loops of chains.

4–2 (continued)

CONTENT DEVELOPMENT

Point out to students that all living things contain polymers. Protein, an essential ingredient of living matter, is a polymer.

Proteins are organic compounds that contain nitrogen in addition to carbon, hydrogen, and oxygen. Proteins are polymers made of monomers called amino acids.

Proteins are found in a variety of substances. Some common proteins and their uses are

collagen—used in tennis racket strings
keratin—found in feathers
silk—from a spider web

● ● ● ● **Integration** ● ● ● ●

Use the discussion of amino acids to integrate physiology into your lesson.

REINFORCEMENT/RETEACHING

▶ *Activity Book*

Students who have difficulty understanding the concepts of this section should be provided with the Chapter 4 activity called A World of Synthetic Polymers.

GUIDED PRACTICE

Skills Development

Skills: Applying concepts, making observations, making comparisons

At this point have students complete the in-text Chapter 4 Laboratory Investigation: Comparing Natural and Synthetic Polymers. In the investigation students will compare the strength, absorbency, and resistance to chemical change of natural and synthetic polymers.

102 ■ O

Polymers produced from petrochemicals are called **synthetic polymers.** Something that is synthetic does not exist naturally. Instead it is made by people. Polymer chemistry has produced synthetic materials that are strong, lightweight, heat resistant, flexible, and durable (long lasting). These properties give polymers a wide range of applications.

Although the term polymer may be new to you, you will soon discover that many polymers produced from petrochemicals are familiar to you. For example, petrochemical products such as synthetic rubber and plastic wrap are synthetic polymers. Synthetic polymers are used to make fabrics such as nylon, rayon, Orlon, and Dacron. Plastics—used in products from kitchen utensils to rocket engines—are petrochemical products made of polymers.

In medicine, polymers are used as substitutes for human tissues, such as bones and arteries. These polymers must last a lifetime and must withstand the wear and tear of constant use. Polymer adhesives, rather than thread, may be used to hold clothes together. Polymers are replacing glass, metal, and paper as containers for food. The cup of hot chocolate you may have held today did not burn your hand because it was made of a white insulating polymer. Polymer materials are also used to make rugs, furniture, wall coverings, and curtains. Look around and see how many polymers you can spot. And remember: You have petroleum to thank for all these useful materials.

Polymer materials also can be mixed and matched to produce substances with unusual properties.

Figure 4–8 *Applications of synthetic polymers include rubber tires, aspirin, and waterproof rain gear.*

ACTIVITY
DOING

Homemade Adhesives

1. Mix flour and water to make a paste.

2. Separate the white from the yolk of a raw egg. Egg white is a natural adhesive.

Compare the "sticking strength" of the flour paste and the egg white by using each to lift objects of increasing mass.

O ■ 103

Figure 4–9 *Different forms of plastics are made by melting and processing small bits of plastic grain. The plastic grain can be made into a variety of products, each with characteristics suited for a specific use.*

Different plastics and synthetic fibers are combined to make puncture-proof tires and bulletproof vests. Layers of polymer materials can be combined to make waterproof rain gear.

Polymer chemistry is also important in the transportation industry. Every year the number of polymer parts in cars, planes, and trains increases. A plastic car engine has been built and tested. This engine is lighter, more fuel efficient, and more durable than a metal engine. As you can see, polymers made from petroleum are extremely important today. And they will be even more important in the future.

Life Science Library/Giant Molecules. Photograph by Donald Miller © 1986 Time, Inc. Time-Life Books, Inc. Publisher

Life Science Library/Giant Molecules. Photograph by John Zimmerman © 1986 Time, Inc. Time-Life Books, Inc. Publisher

Figure 4–10 *Polymer technology has produced some amazing materials. This sheet of plastic bends no matter how hard it is struck by a hammer (bottom). This extra-thin sheet of plastic is not damaged by temperatures greater than 1000°C (top).*

104 ■ O

4–2 Section Review

1. What is a petrochemical product? Give several examples.
2. What is the relationship between a monomer and a polymer?
3. List three examples of natural polymers.
4. What is polymerization?
5. What are some characteristics of synthetic polymers?

Connection—*You and Your World*
6. What might be some of the economic side effects of increased use of polymers in automobiles?

CONNECTIONS

Life-Saving Chemistry ❶

The human body is an amazing collection of systems that interact with one another to sustain life. The systems are made up of tissues and cells that each have a certain function. For example, thin, soft, flexible layers of tissue control the flow of chemicals into and out of the cells. These layers are called *biological membranes*. Some biological membranes allow harmful chemicals to pass out of a cell while keeping needed chemicals inside the cell. The kidneys contain membranes that do this kind of job. Biological membranes are natural polymers.

Like other parts of the human body, biological membranes sometimes do not work correctly or become damaged. The failure of a biological membrane to perform its necessary function can be fatal. That is why healthy kidneys are so important. In the human body, two kidneys constantly filter waste materials from the blood. If the kidneys fail to do this vital job, wastes will remain in the blood, causing damage to other parts of the body.

In 1944, a synthetic membrane that worked like a natural biological membrane

was developed. This synthetic membrane made possible the invention of the artificial kidney. The artificial kidney is essentially a large filtering machine. A tube from the machine is inserted into an artery in the patient's arm. Blood flows from the patient into the machine, where the blood is filtered by a synthetic membrane. The purified blood is then pumped back into the patient's system.

Synthetic membranes have a variety of other applications. Some are used as skin substitutes for people who have been badly burned. Others are used in artificial body parts. Polymer chemistry has had a significant impact on human health.

No, the bird is not drowning. It is surrounded by a synthetic membrane that allows oxygen to pass from the water to the bird.

ANNOTATION KEY

Integration
❶ Life Science: Excretory System. See *Human Biology and Health*, Chapter 6.

CONNECTIONS
LIFE-SAVING CHEMISTRY

Remind students that a functioning kidney is absolutely essential to life. Fortunately, most of us have built-in insurance against a threat to our survival because of kidney damage. We have two kidneys—one more than is necessary to live a healthy, normal life. Until recently, people who developed severe kidney disease and lost both kidneys could not survive for more than a few months or weeks. Now, however, renal dialysis machines can mimic the functions of the human kidney and can sustain life until a kidney transplant can be arranged. One problem that doctors face when transplanting donated or artificial kidneys is that the patient's immune system may reject the foreign material.

If you are teaching thematically, you may want to use the Connections feature to reinforce the themes of systems and interactions, and stability.

Integration: Use the Connections feature to integrate the excretory system into your science lesson.

5. Synthetic polymers can be strong, lightweight, heat resistant, flexible, and durable.

6. Answers may vary. Students may suggest that the use of polymers may result in more efficient cars, safer cars, or more expensive cars.

REINFORCEMENT/RETEACHING

Review students' responses to the Section Review questions. Reteach any material that is still unclear, based on students' responses.

CLOSURE

▶ *Review and Reinforcement Guide*

Have students complete Section 4–2 in the *Review and Reinforcement Guide*.

GUIDED PRACTICE

▶ *Laboratory Manual*

Skills Development

Skills: Applying concepts, making observations, making comparisons

At this point you may want to have students complete the Chapter 4 Laboratory Investigation in the *Laboratory Manual* called Tensile Strength of Natural and Synthetic Polymers. In the investigation students will test the tensile strength of threads made of cotton and polyester.

INDEPENDENT PRACTICE

Section Review 4–2

1. A petrochemical product, such as plastic or synthetic rubber, is a product made from petroleum.

2. Polymers are made up of chains of monomers.

3. Cotton, silk, and wool are all natural polymers.

4. Polymerization is the process of chemically bonding monomers to form polymers.

Laboratory Investigation

COMPARING NATURAL AND SYNTHETIC POLYMERS

BEFORE THE LAB

1. Gather all materials at least one day prior to the investigation. You should have enough supplies to meet your class needs, assuming six students per group.
2. If you have difficulty obtaining linen, silk can be used instead. If you have difficulty obtaining acetate, acrylic fiber can be substituted.

PRE-LAB DISCUSSION

Begin the discussion by asking students,
• **Why do you think synthetic fabrics were invented?** (Answers may vary. Probably, synthetic fabrics were invented in order to obtain fabrics with certain properties, such as water repellency or the ability to "wash and wear" with little or no ironing. Also, it is possible that at certain times in certain places, natural fabrics may have been in short supply or difficult or expensive to obtain.)

Have students read the complete laboratory procedure. Discuss the procedure by asking questions similar to the following:
• **What is being investigated in this lab?** (The properties of natural vs. synthetic fibers.)
• **What is the variable in this investigation?** (The different types of fabric and whether the fabric is made from natural or synthetic polymers.)
• **What is the control?** (The things that will be done to the fabrics.)
• **What is your hypothesis as to the outcome of the investigation?** (Answers will vary, but they should reflect an expectation to see a difference in properties between synthetic and natural fibers.)

SAFETY TIPS

Remind students to wear safety goggles and rubber gloves during the lab. Caution them to avoid splashing bleach on skin or clothing. At the end of the lab, emphasize the need to dispose of liquid wastes properly.

Laboratory Investigation

Comparing Natural and Synthetic Polymers

Problem

How do natural and synthetic polymers compare in strength, absorbency, and resistance to chemical damage?

Materials *(per group)*

3 samples of natural polymer cloth:
    wool, cotton, linen
3 samples of synthetic polymer cloth:
    polyester, nylon, acetate
12 Styrofoam cups
mild acid (lime or lemon juice or vinegar)

marking pen	liquid bleach
metric ruler	medicine dropper
scissors	oil
rubber gloves	paper towel

Procedure 🧪 📷 👁 📏

1. Record the color of each cloth.
2. Label 6 Styrofoam cups with the names of the 6 cloth samples. Also write the word Bleach on each cup.
3. Cut a 2-square-cm piece from each cloth. Put each piece in its cup.
4. Wearing rubber gloves, carefully pour a small amount of bleach into each cup.
5. Label the 6 remaining cups with the names of the 6 cloth samples and the word Acid. Then pour a small amount of the mild acid into each and repeat step 3.

6. Set the cups aside for 24 hours. Meanwhile, proceed with steps 7 through 9.
7. Using the remaining samples of cloth, attempt to tear each.
8. Place a drop of water on each material. Note whether the water forms beads or is absorbed. If the water is absorbed, record the rate of absorption.
9. Repeat step 8 using a drop of oil.
10. After 24 hours, wearing rubber gloves, carefully pour the liquids in the cups into the sink or a container provided by your teacher. Dry the samples with a paper towel.
11. Record any color changes.

Observations

1. Which material held its color best in bleach? In acid?
2. Which materials were least resistant to chemical damage by bleach or mild acid?
3. Which material has the strongest fiber or is hardest to tear?
4. Which materials are water repellent?

Analysis and Conclusions

1. Compare the natural and synthetic polymers' strength, absorbency, and resistance to chemical change.
2. Which material would you use to manufacture a laboratory coat? A farmer's overalls? A raincoat? An auto mechanic's shirt?
3. **On Your Own** Confirm your results with additional samples of natural and synthetic polymers. What additional tests can you add for comparison?

TEACHING STRATEGY

1. Have teams follow the directions carefully as they work in the laboratory.
2. Suggest that students put aside a sample of each polymer cloth to use in comparing the before/after characteristics.
3. Suggest that students look for other fabric changes besides color during the activity.
4. You may want to distribute magnifying glasses for students to observe small differences in the polymer fibers when comparing the product of each test to the untreated polymer.

DISCOVERY STRATEGIES

Discuss how the investigation relates to the chapter ideas by asking open questions similar to the following:
• **Why did we set aside a piece of each polymer cloth?** (To use it as a control for comparing before/after each treatment or test—making comparisons.)
• **What were the experimental variables?** (Bleach, water, acid, tearing, and oil—making observations.)
• **What other variables might be tested to find out how the polymers react?** (Stu-

Study Guide

Summarizing Key Concepts

4–1 What Is Petroleum?

▲ Petroleum is a substance believed to have been formed when plant and animal remains were subjected to tremendous pressure for millions of years.

▲ Petroleum is a mixture of chemicals that can be divided into separate parts, or fractions.

▲ Petroleum is separated into its components through a process call fractional distillation.

▲ During fractional distillation, petroleum is heated and pumped into a fractionating tower. As the fractions rise up the tower, they cool and condense. Because they condense at different temperatures, they can be drawn off at different heights.

▲ Different groups of products are produced from petroleum: raw materials used in manufacturing, raw materials converted to chemicals, lubricants, and fuels.

4–2 Petrochemical Products

▲ Substances derived from petroleum are called petrochemical products.

▲ Most petrochemical products are polymers.

▲ A polymer is a series of molecular units called monomers.

▲ The process of chemically combining monomers to make a polymer is called polymerization.

▲ Natural polymers include cotton, silk, wool, natural rubber, cellulose, protein, and lignin.

▲ Synthetic polymers include synthetic rubber, plastics, and fabrics such as nylon, Orlon, rayon, and Dacron.

▲ Polymers are usually strong, lightweight, heat resistant, flexible, and durable.

▲ Polymers can be mixed and matched to form substances that are waterproof, puncture proof, or electrically conductive.

Reviewing Key Terms

Define each term in a complete sentence.

4–1 What Is Petroleum?
petroleum
fraction
refining

4–2 Petrochemical Products
petrochemical product
monomer
polymer
natural polymer
polymerization
synthetic polymer

ANALYSIS AND CONCLUSIONS

1. The synthetic polymer polyester was the strongest in terms of being hardest to tear. The synthetic polyester absorbed the least amount of water, whereas the synthetic polymers polyester and nylon absorbed the least amount of oil. The natural polymer wool was the most resistant to chemical wear from bleach and acid.

2. Lab coat: cotton, linen; farmer's overalls: polyester; raincoat: polyester, nylon; auto mechanic's shirt: polyester, nylon.

3. Students may suggest testing the fabric samples for fire resistance or stain resistance.

GOING FURTHER: ENRICHMENT

Part 1

Have students perform a similar experiment using synthetic rubber and natural rubber or plastic and wood. Encourage the students to compare the usefulness of items made from these substances, based on the properties they discover.

Part 2

Have students take an inventory of their clothing and record five items made from synthetic fibers and five items made from natural fibers. Have them write a brief statement describing how each garment has worn so far, noting any problems or benefits. For example, a student might write, "Dacron blouse: Needs no ironing after it is washed, but color has not stayed as bright as I had hoped." Have students share and compare their lists in a class discussion.

dents might suggest boiling, using dyes, washing in strong detergent, drying in a dryer, making the cloth soiled with various substances and then washing, and testing to find out about wrinkle and crease resistance or retention—making comparisons.)

OBSERVATIONS

1. Wool, wool.
2. Cotton, linen.
3. Polyester.
4. Polyester, nylon.

Chapter Review

ALTERNATIVE ASSESSMENT

The *Prentice Hall Science* program includes a variety of testing components and methodologies. Aside from the Chapter Review questions, you may opt to use the Chapter Test or the Computer Test Bank Test in your *Test Book* for assessment of important facts and concepts. In addition, Performance-Based Tests are included in your *Test Book*. These Performance-Based Tests are designed to test science process skills, rather than factual content recall. Since they are not content dependent, Performance-Based Tests can be distributed after students complete a chapter or after they complete the entire textbook.

CONTENT REVIEW

Multiple Choice

1. d
2. b
3. a
4. c
5. a
6. b
7. a
8. c
9. c
10. d

True or False

1. F, crude oil
2. T
3. F, fractional distillation
4. T
5. F, synthetic
6. F, condenses
7. F, polymer, monomers
8. T

Concept Mapping

Row 1: Fractions, Substance
Row 2: Fractional distillation, buried animals and plants

CONCEPT MASTERY

1. Plants and animals died out and were buried by layers of sediment. Over millions of years, pressure and heat transformed the organic material into petroleum.
2. The different fractions of petroleum have different boiling points. The petroleum is heated to a high temperature and then piped into a fractionating tower. The different fractions condense at different temperatures and heights in the tower and are drawn off as they condense.
3. They condense near the bottom of the tower or do not condense at all. They condense higher up the tower.
4. It is a source of energy that can easily be used for heating and transportation.
5. Naturally occurring substances with a particular molecular structure. Examples include cotton, silk, wool, and natural rubber.
6. Polymerization is the process of chemically bonding monomers to form polymers. It can be used to produce synthetic materials with desirable properties.
7. Synthetic polymers are substances produced from petrochemicals that have a particular molecular structure. Examples include nylon, rayon, plastics, and synthetic rubber.
8. They are strong, light, heat resistant, flexible, and long lasting.

Chapter Review

Content Review

Multiple Choice

Choose the letter of the answer that best completes each statement.

1. Crude oil is
   a. a single element.
   b. gasoline.
   c. asphalt.
   d. a mixture.
2. The physical property used to separate petroleum into its parts is
   a. melting point.
   b. boiling point.
   c. density.
   d. solubility.
3. The highest temperature in the fractionating tower is
   a. below the boiling point of most petroleum fractions.
   b. above the boiling point of most petroleum fractions.
   c. equal to the boiling point of most petroleum fractions.
   d. below the melting point of most petroleum fractions.
4. A substance unlikely to vaporize in a fractionating tower is
   a. kerosene.
   b. gasoline.
   c. asphalt.
   d. heating fuel.
5. The process of distillation involves
   a. vaporization and condensation.
   b. freezing and melting.
   c. vaporization and melting.
   d. freezing and condensation.
6. A polymer is made of a series of
   a. atoms.
   b. monomers.
   c. fuels.
   d. synthetic molecules.
7. An example of a natural polymer is
   a. wool.
   b. plastic.
   c. crude oil.
   d. copper.
8. An example of a synthetic polymer is
   a. natural rubber.
   b. protein.
   c. rayon.
   d. cotton.
9. The process of chemically bonding monomers to form polymers is called
   a. distillation.
   b. fractionation.
   c. polymerization.
   d. refining.
10. An example of a polymer product is
    a. lead tubing.
    b. crude oil.
    c. water.
    d. plastic.

True or False

If the statement is true, write "true." If it is false, change the underlined word or words to make the statement true.

1. Petroleum taken directly from the Earth is called <u>asphalt</u>.
2. Petroleum can be separated into its different parts, or <u>fractions</u>.
3. The process of separating petroleum into its components is called <u>condensing</u>.
4. <u>Polymerization</u> involves the chemical bonding of monomers into polymers.
5. Plastics are examples of <u>natural</u> polymers.
6. When a vapor <u>evaporates</u>, it changes back to a liquid.
7. A <u>monomer</u> is a long chain of <u>polymers</u>.
8. Silk is an example of a <u>natural</u> polymer.

Concept Mapping

Complete the following concept map for Section 4–1. Refer to pages O6–O7 to construct a concept map for the entire chapter.

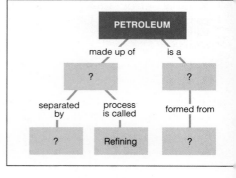

108 ■ O

Concept Mastery

Discuss each of the following in a brief paragraph.

1. How did petroleum most likely form?
2. Explain how petroleum is separated into parts during fractional distillation.
3. What happens in the fractionating tower to fractions with extremely high boiling points? With extremely low boiling points?
4. Why is petroleum an important source of fuel?
5. What are natural polymers? Give some examples.
6. What is polymerization? Why is this process useful?
7. What are synthetic polymers? Give some examples.
8. What characteristics of synthetic polymers make them so useful?

Critical Thinking and Problem Solving

Use the skills you have developed in this chapter to answer each of the following.

1. **Drawing conclusions** Petroleum is believed to have been formed at the bottom of oceans. However, oil is often found under dry land—even in deserts. How can you explain this?

2. **Relating facts** Describe some of the uses of synthetic polymers in your life. Then describe what changes you would have to make in your lifestyle if these polymers were not available.
3. **Identifying patterns** The achievements and inventions of the United States space program—originally directed outside planet Earth—have found dramatic uses in our everyday lives. Why do you think space-technology spinoffs have been so widely and successfully used?
4. **Applying concepts** Describe several ways in which polymer chemistry can be used in medicine other than those mentioned in the chapter.
5. **Analyzing information** The United States has been said to have undergone a chemical revolution during the last fifty years. Having read this chapter, explain what the term chemical revolution means to you.
6. **Applying technology** Imagine that you are an engineer whose task is to design a fuel-efficient car that meets all the current standards for safety and durabilty. What kinds of materials would you consider using? What properties must the materials used in the engine have? How about the materials used in safety belts and seat cushions?
7. **Using the writing process** Many people use and enjoy the products derived from petroleum. However, some of the chemical processes used to make these products—as well as the burning of petroleum fuels—add to the pollution of the air, land, and water. Should the manufacturing of certain products be stopped in order to protect the environment? Write a composition expressing your opinion and explaining the reason for it.

---

tures that build up in a car engine. Furthermore, materials used in engines should be lightweight in order to conserve fuel. Materials used in safety belts should be flexible but very strong and able to withstand the stresses involved during an accident. The seat cushions should be comfortable, durable, and easily cleaned. Students might also include materials for air bags in their responses.

7. There are no right or wrong answers to this issue. Some students will show great concern for the environment and may suggest cutting back on our use of petroleum. Others may feel that the need for petroleum products outweighs the pollution challenges. Still others may suggest finding methods to decrease the environmental impact of our use of petroleum products.

KEEPING A PORTFOLIO

You might want to assign some of the Concept Mastery and Critical Thinking and Problem Solving questions as homework and have students include their responses to unassigned questions in their portfolio. Students should be encouraged to include both the question and the answer in their portfolio.

ISSUES IN SCIENCE

The following issues can be used as springboards for discussion or given as writing assignments:

1. The proposed construction of the Alaska Pipeline in the early 1970s prompted tremendous criticism from conservationists, who were able to block the pipeline's construction for several years. Find out why the conservationists objected and why they were eventually overruled. Then express your own opinion on the issue.

2. Despite the popularity of synthetic fabrics, many people will not buy a garment unless it is marked 100 percent natural fiber. They feel that synthetic fibers "don't breathe," are not good for their skin, and are "not organic." Take a position on this issue and explain why you will or will not wear synthetic fabrics.

CRITICAL THINKING AND PROBLEM SOLVING

1. The dry land may have been under an ocean at some time in the past.

2. Answers will vary, depending on students' selections. Check that students' answers are logical and well written.

3. Students' responses might include the fact that many polymers have been developed for the space program. These would include materials that must resist high temperatures, pressure, and heat.

4. The textbook chapter mentions substitutes for bones and tissues such as human arteries. Students may suggest other prosthetic devices, artificial kidneys, or clothing and protective equipment used in hospitals.

5. Students' essays will vary but should include the petrochemical products mentioned in the chapter.

6. Students should point out that materials used in an engine must be durable, must be able to withstand vibrations, and must be able to withstand high tempera-

Chapter 5   RADIOACTIVE ELEMENTS

SECTION	HANDS-ON ACTIVITIES
5–1 Radioactivity pages O112–O115 Multicultural Opportunity 5–1, p. O112 ESL Strategy 5–1, p. O112	**Activity Book** CHAPTER DISCOVERY: Modeling a Decay Series, p. O105
5–2 Nuclear Reactions pages O116–O125 Multicultural Opportunity 5–2, p. O116 ESL Strategy 5–2, p. O116	**Student Edition** LABORATORY INVESTIGATION: The Half-Life of a Sugar Cube, p. O136 **Laboratory Manual** Half-Life of a Capacitor, p. O61
5–3 Harnessing the Nucleus pages O126–O130 Multicultural Opportunity 5–3, p. O126 ESL Strategy 5–3, p. O126	**Student Edition** ACTIVITY BANK: The Domino Effect, p. 159 **Teacher Edition** Linking Up With Science, p. O110d
5–4 Detecting and Using Radioactivity pages O131–O135 Multicultural Opportunity 5– 4, p. O131 ESL Strategy 5– 4, p. O131	**Student Edition** ACTIVITY (Doing): Radioactivity in Medicine, p. O133 **Teacher Edition** Exploring Science Dating, p. O110d
Chapter Review pages O136–O139	

OUTSIDE TEACHER RESOURCES

Books

Condon, E. V., and H. Odabasi. *Atomic Structure,* Cambridge University Press.

Conn, G. K. *Atoms and Their Structure,* Cambridge University Press.

Hagel, J., III. *Alternative Energy Strategies: Constraints and Opportunities,* Praeger.

McGowen, Tom, *Radioactivity: From the Curies to the Atomic Age,* Watts.

Pringle, Laurence. *Nuclear Energy: Troubled Past, Uncertain Future,* Macmillan.

OTHER ACTIVITIES	MEDIA AND TECHNOLOGY
Activity Book ACTIVITY: Alpha, Beta, and Gamma Radiation in a Magnetic Field, p. O109 **Review and Reinforcement Guide** Section 5–1, p. O45	**English/Spanish Audiotapes** Section 5–1
Student Edition ACTIVITY (Writing): What Is a Quark? p. O117 ACTIVITY (Writing): The Nuclear Age, p. O118 ACTIVITY (Thinking): Interpreting Nuclear Equations, p. O124 **Activity Book** ACTIVITY: Decay Series of Uranium-238, p. O113 ACTIVITY: Radioactive and Nonradioactive Elements at a Glance, p. O107 ACTIVITY: Atomic Arithmetic, p. O115 ACTIVITY: Nuclear Equations, p. O117 **Review and Reinforcement Guide** Section 5–2, p. O47	**English/Spanish Audiotapes** Section 5–2
Student Edition ACTIVITY (Reading): An Event That Would Change History Forever, p. O127 **Activity Book** ACTIVITY: Nuclear Fission and Nuclear Fusion, p. O111 **Review and Reinforcement Guide** Section 5–3, p. O51	**English/Spanish Audiotapes** Section 5–3
Review and Reinforcement Guide Section 5–4, p. O53	**English/Spanish Audiotapes** Section 5–4
Test Book Chapter Test, p. O99 Performance-Based Tests, p. O117	**Test Book** Computer Test Bank Test, p. O105

*All materials in the Chapter Planning Guide Grid are available as part of the Prentice Hall Science Learning System.

Audiovisuals

The Atomic Nucleus, coarseware, Prentice-Hall

Matter and Molecules: Into the Atom, filmstrip, Singer Educational Division

Measuring Electron Charge and Mass, filmstrips, Prentice-Hall Media

The Nucleus: Composition, Stability, and Decay, filmstrip or slides, Prentice-Hall Media

CHAPTER OVERVIEW

The discovery of radiation by Henri Becquerel in 1896 provided a stimulus for other researchers and scientists to develop new technologies and uses of radioactivity.

Radioactivity is a naturally occurring phenomenon—radioactive elements are found naturally within the Earth. Both natural and unnaturally occurring radioactive elements and isotopes have many uses, especially in the medical treatment and diagnosis of disease.

Radioactivity can also be artificially initiated. It is used today for the creation of new elements and as a source of energy such as electric power. Artificially initiated nuclear reactions have been packaged as war weapons capable of massive destruction.

5–1 RADIOACTIVITY

THEMATIC FOCUS

The purpose of this section is to introduce students to the concept of radioactivity. Students will learn about the work of early researchers such as Henri Becquerel and Marie and Pierre Curie, and how their work led to an understanding of the nature of nuclear radiation. Today, scientists use the term *radioactivity* to describe the release of nuclear radiation in the form of particles and rays from a radioactive element. Students will discover that the three types of radiation, or decay, given off by radioactive elements have been named alpha particles, beta particles, and gamma rays, after the first three letters of the Greek alphabet. It will be revealed that the alpha particles are weakest in terms of strength—they can be stopped by a piece of paper; that beta particles can pass through approximately 3 millimeters of aluminum; and that gamma rays can penetrate through several centimeters of lead.

The theme that can be focused on in this section is unity and diversity.

Unity and diversity: Radioactive decay involves the emission of fragments of the nucleus.

PERFORMANCE OBJECTIVES 5–1

1. Define radioactivity.
2. Distinguish between alpha particles, beta particles, and gamma rays.

SCIENCE TERMS 5–1

nuclear radiation p. O113
radioactive p. O113
radioactivity p. O114
alpha particle p. O115
beta particle p. O115
gamma ray p. O115

5–2 NUCLEAR REACTIONS

THEMATIC FOCUS

This section introduces students to nuclear stability, radioactive decay, radioactive half-life, and artificial transmutation.

Students will learn that the nucleus of an atom is held together by a force known as the strong force and with energy known as binding energy. Binding energy is essential to the stability of a nucleus—too little binding energy allows a nucleus to be unstable. These unstable nuclei can become stable by undergoing a nuclear reaction, or change. One of the four ways that nuclei can stabilize is through decay. During decay, a nucleus emits radioactive energy in the form of alpha particles, beta particles, or gamma rays. Alpha and beta decay is usually accompanied by gamma emission.

Any radioactive element decays at a predictable rate. This rate of decay is known as the half-life of the element. Knowledge of half-lives allows scientists to date fossils or objects that may be thousands or millions of years old.

A nuclear reaction known as artificial transmutation allows scientists to decay elements artificially and to create new elements in the process.

The themes that can be focused on in this section are energy, patterns of change, systems and interactions, stability, evolution, and unity and diversity.

*Energy: Binding energy provided by the nuclear strong force holds the nucleus together. If binding energy is low, a nucleus will be radioactive.

*Patterns of change: If an unstable nucleus can become stable by changing, it will undergo a nuclear reaction.

*Systems and interactions: Bombarding the nucleus with subatomic particles can result in transmutation.

*Stability: An element that undergoes radioactive decay will continue to decay until a stable nucleus is achieved. The various elements produced in the process constitute a decay series.

Evolution: The age of fossils can be determined using the half-lives of certain radioactive elements.

Unity and diversity: Artificial transmutation involves the emission of fragments of the nucleus.

PERFORMANCE OBJECTIVES 5–2

1. Define binding energy.
2. Compare radioactive and stable isotopes.
3. Describe and contrast alpha, beta, and gamma decay.
4. Explain and apply the concept of half-life.
5. Define transmutation.
6. Explain artificial transmutation.

SCIENCE TERMS 5–2

nuclear strong force p. O116
binding energy p. O116
isotope p. O116
radioactive decay p. O118
transmutation p. O119
half-life p. O120
decay series p. O121
artificial transmutation p. O122
transuranium element p. O123

5–3 HARNESSING THE NUCLEUS
THEMATIC FOCUS

Students will learn about the processes of nuclear fission and nuclear fusion in this section. Nuclear fission involves the splitting of an atomic nucleus into two smaller nuclei of approximately equal mass. Much of the early development of fission occurred in the 1930s. A typical fission reaction involves accelerating a particle, or "bullet," to a speed almost equal to the speed of light. The particle is then allowed to collide with an atom, violently smashing the atom into two pieces. The amount of energy released when an atom splits in this fashion is rather insignificant. It was discovered, however, that if the reaction could be repeated over and over again, tremendous amounts of energy would be released. The neutrons released by a splitting atom act as additional "bullets" in a fission reaction, causing more and more atoms to be split. This uncontrolled reaction releases huge amounts of energy, and scientists have learned to control this reaction and to build power plants that use fission energy to help generate electricity.

The process of fusion involves the joining of two atomic nuclei of smaller masses to form a single nucleus of larger mass. Fusion requires extreme pressure and temperature; as a result, its duplication is extremely difficult. Fusion, however, offers many advantages when compared to nuclear fission, if time and technology allow its duplication.

The themes that can be focused on in this section are energy, scale and structure, and unity and diversity.

*Energy: Tremendous amounts of energy are released during fission and fusion.

*Scale and structure: One individual fission reaction releases a relatively small amount of energy. The neutrons emitted during the reaction, however, cause other nuclei to undergo fission in a chain reaction. The total amount of energy released is tremendous.

Unity and diversity: Fission involves splitting a nucleus into two. Fusion involves combining two nuclei to form one.

PERFORMANCE OBJECTIVES 5–3
1. Distinguish between the processes of nuclear fission and nuclear fusion.
2. Trace the events in a nuclear chain reaction.

SCIENCE TERMS 5–3
nuclear fission p. O126
nuclear chain reaction p. O127
nuclear fusion p. O128

5–4 DETECTING AND USING RADIOACTIVITY
THEMATIC FOCUS

In this section students are introduced to various instruments that scientists use to detect and measure radioactivity. These instruments include the electroscope, the Geiger counter, the cloud chamber, and the bubble chamber.

Various practical uses of radioisotopes are also described. These uses include disease diagnosis and treatment, food sterilization, and use as tracers. The dangers of using, working with, or coming in contact with radioactive substances are also described.

The theme that can be focused on in this section is patterns of change.

*Patterns of change: The development of nuclear technologies changes and enhances the quality of life.

PERFORMANCE OBJECTIVES 5–4
1. Compare and contrast the use of Geiger counters, electroscopes, cloud chambers, and bubble chambers.
2. Describe the use of radioisotopes in diagnosing and treating disease and in other applications.
3. Describe the dangers associated with radioactive substances and the precautions taken to minimize the dangers.

SCIENCE TERMS 5–4
electroscope p. O131
Geiger counter p. O131
cloud chamber p. O132
bubble chamber p. O132
radioisotope p. O133
tracer p. O133

Discovery Learning

TEACHER DEMONSTRATIONS MODELING
Linking Up With Science

Show the class a chain. Use a clamp to attach one end of the chain to a table. Pull on the free end of the chain and ask students to observe the results.

• **What happened to each link in the chain as the end of the chain was pulled?** (Each link was pulled by the link before it. The force and energy were transferred from link to link.)

Set two toy cars on the demonstration table. Have them face in the same direction with their bumpers touching. Use a third car to crash into the back of the rear car.

• **What happened?** (Lead students to use the term *chain reaction.*)

Point out that chain reactions occur in nuclear fission reactions.

Exploring Science Dating

Obtain a very old bone from a biology instructor, an artifact such as an old piece of pottery from a social studies instructor, or an arrowhead from a collection. Display the object for students to observe.

• **How old do you guess this might be?** (Accept all predictions.)

• **How could you discover its age?** (Students might suggest performing library research and so forth.)

Point out that determining the age of fossils or other objects has been made easier because of the knowledge gained from nuclear research and that students will learn more about this kind of dating in their studies of this section of their textbook.

CHAPTER 5
Radioactive Elements

INTEGRATING SCIENCE

This physical science chapter provides you with numerous opportunities to integrate other areas of science, as well as other disciplines, into your curriculum. Blue numbered annotations on the student page and integration notes on the teacher wraparound pages alert you to areas of possible integration.

In this chapter you can integrate social studies (p. 112), physical science and electromagnetic radiation (p. 113), language arts (pp. 113, 114, 117, 118, 119, 127, 128, 133), physical science and atomic structure (p. 115), life science and radioactive dating (p. 117), mathematics (p. 120), life science and evolution (p. 121), earth science and astronomy (p. 124), earth science and ecology (p. 125), physical science and nuclear fission (p. 126), physical science and nuclear fusion (p. 129), physical science and charge (p. 131), life science and botany (p. 132), life science and medicine (p. 133), physical science and MRI (p. 134), and life science and genetics (p. 134).

SCIENCE, TECHNOLOGY, AND SOCIETY/COOPERATIVE LEARNING

Radioactive elements have given us nuclear power, a reliable dating method, and a major problem—how do we dispose of radioactive waste material? Radioactive waste material can remain harmful to living things for up to 3 million years!

Scientists have been searching for the perfect geological site for radioactive waste for years. Many of the sites they have selected have met with local opposition that has effectively stalemated the selection of a permanent nuclear graveyard for our country's radioactive waste.

The site selected must isolate the waste for at least 10,000 years, be geologically safe, and be far enough from groundwater to prevent contamination. In the repository, a combination of artificial and natural barriers will be used to isolate the waste. The waste will be buried in stainless-steel canisters placed in storage shafts

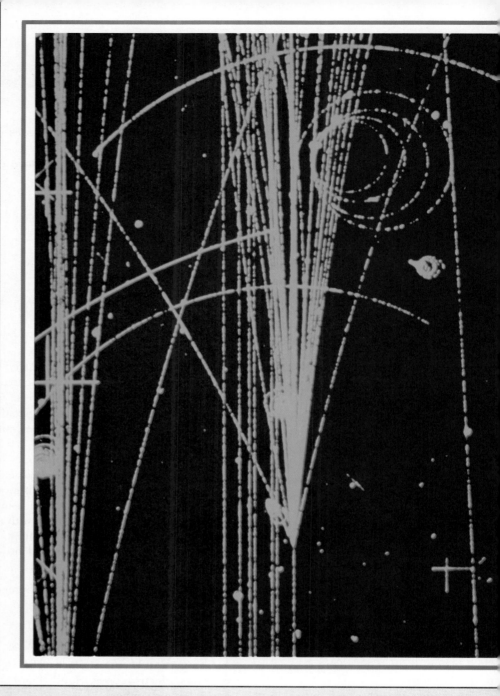

INTRODUCING CHAPTER 5

DISCOVERY LEARNING

▶ *Activity Book*

Begin your teaching of the chapter by using the Chapter 5 Discover Activity from the *Activity Book*. Using this activity, students will create a model of a portion of the decay series for uranium-238.

USING THE TEXTBOOK

Have students observe the picture on page O110 and read its caption.

Point out that the tracks or "map" they are observing is an essential tool of scientists studying the atom. Explain that these strange markings are produced when subatomic particles, such as protons, are accelerated to great speeds and are made to collide with one another. During the collisions, energy and new particles are given off. The "map" actually

Radioactive Elements

The rich prairie land near Waxahachie, Texas, is the site of one of the most ambitious scientific research projects ever undertaken. The project will take many years and billions of dollars to complete. Yet even when construction ends, it will appear that not much has been done. Why? This project is being built almost 46 meters beneath the soil.

The project is the world's largest accelerator, known as the Superconducting Supercollider (SSC). The supercollider will be housed in a circular tunnel some 85 kilometers in length! Protons will shoot around and around the tunnel, controlled by more than 9000 superconducting magnets that will keep the protons on track and help boost their speed.

When the protons are traveling at almost the speed of light, they will be diverted toward the nuclei of target atoms. The reactions that occur during the collisions of protons and target nuclei hold long sought-after information.

Why is smashing tiny particles into atomic nuclei worth all the effort? In this chapter you will discover the importance of subatomic particles, their relation to the nucleus of an atom, and their role in producing enormous quantities of energy.

Journal *Activity*

You and Your World You have probably heard the terms atomic bomb (A-bomb) and hydrogen bomb (H-bomb). Perhaps you even know about the nuclear radiation associated with them. In your journal, describe the thoughts and/or questions that come to mind when you hear these terms.

Studying the tracks made by subatomic particles enables scientists to learn more about matter and energy.

O ■ 111

---

deep below the ground and protected by layers and layers of rock, and each shaft will be cemented shut.

Once the perfect nuclear graveyard is identified and threats posed by natural disasters have been controlled by engineering, the remaining challenge is the curiosity of humans!

Cooperative learning: Using preassigned lab groups or randomly selected teams, have groups complete the following assignment:
- Each group has been selected to design a nuclear hazard warning system for the United States. Because signs can rust away, buildings can collapse, and languages can change, students must create a symbolic language for their warning system that will resist change for 10,000 years and will be understandable to all cultures. You might want to suggest to the groups that they begin by generating a list of words that they want to use in communicating hazards and then design several different symbols for each word. As a group, they can then select the symbols that best communicate the word. Each group should also produce a warning poster using their symbolic language. After displaying each group's poster, have each group work together to translate the posters produced by the other groups.

See Cooperative Learning in the *Teacher's Desk Reference.*

JOURNAL ACTIVITY

You may want to use the Journal Activity as the basis of a class discussion. Given the recent political changes throughout the world, have students discuss what they think the likelihood of a nuclear war might be. Regardless of their opinion, have students support their position. You might also choose to have students relate their opinions on a numerical scale—for example, a scale of 1 to 10 with 1 representing impossible and 10 representing certain. Students should be instructed to keep their Journal Activity in their portfolio.

---

shows the collisions and the emission of new particles after the collisions.

Have students read the introductory material on page O111. Emphasize that the experiment described was carried out in a deep underground laboratory. The experiment was a search for the detection of proton decay. For many years, scientists believed that the proton was stable and could not decay. New theories, however, raise the possibility of proton decay. While discussing this experiment, help students develop an appreciation and a realization of the significance of the experiments. In doing so, avoid the introduction of specific terms at this time, but do point out that the theories being tested apply to our most basic understanding regarding the structure of matter and the origins and ultimate fate of our universe.

5–1 Radioactivity

MULTICULTURAL OPPORTUNITY 5–1

Have students investigate the life and work of Marie Curie. Dr. Curie was a pioneer in the discovery of radioactivity and the first woman to receive a Nobel prize for chemistry.

ESL STRATEGY 5–1

The Intruder: Have students circle the term that does not belong in the group below and name the group these terms represent. Then have students write definitions for the uncircled terms.

• alpha particles, zeta particles, gamma rays, beta particles

Figure 5–1 *The image on the photographic film (right) convinced Becquerel that an invisible "something" had been given off by uranium. The photograph on the left shows rectangular blocks containing the element cesium. The photograph was taken through a heavy glass window 1 meter thick with no source of illumination other than the cesium.*

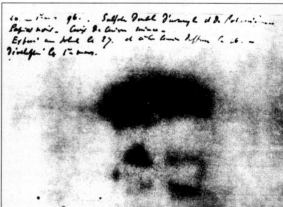

5–1 Radioactivity

Some discoveries are made by performing experiments to find out whether hypotheses are true. Other discoveries are stumbled upon purely by accident. The majority of scientific discoveries, however, are a combination of the two—both genius and luck. One such discovery was made by the French scientist Henri Becquerel (beh-KREHL) in 1896. Becquerel was experimenting with a uranium compound to determine whether it gave off X-rays. His experiments did indeed provide evidence of X-rays. But they also showed something else—something rather exciting. Quite by accident, Becquerel discovered that the uranium compound gave off other types of rays that had never before been detected. Little did Becquerel know then that these mysterious rays would open up a whole new world of modern science.

An Illuminating Discovery

At the time of Becquerel's work, scientists knew that certain substances glowed when exposed to sunlight. Such substances are said to be fluorescent. Becquerel wondered whether in addition to glowing fluorescent substances gave off X-rays.

To test his hypothesis, Becquerel wrapped some photographic film in lightproof paper (paper that does not allow light through it). He placed a piece

TEACHING STRATEGY 5–1

FOCUS/MOTIVATION

Direct students' attention to Figure 5–1, which shows a piece of photographic film exposed by radiation from uranium. Draw upon students' experience and familiarity with cameras and film. Ask:

• **What causes film to develop images when it is exposed?** (Light causes silver bromide to decompose into black metallic silver.)

• **What is light?** (Light is a form of electromagnetic energy.)

• **What does the exposure of film by uranium demonstrate about uranium?** (It is releasing energy.)

CONTENT DEVELOPMENT

As you discuss and students have read about Becquerel's discovery of radiation, ask students to point out how the discovery and the reasoning involved the scientific method. Ask:

• **What was Becquerel's hypothesis?** (He wanted to know if a uranium compound gave off X-rays.)

• **Did Becquerel's hypothesis prove correct?** (Partially. The compound released energy, but not in the form of X-rays.)

Remind students that the discovery was "accidental."

• **Name several "accidental" discoveries you have made in your life.** (Accept all discoveries. Students might suggest, for example, that they discovered a new way to solve a mathematics problem.)

• **Was Becquerel's accidental discovery enough evidence to say with certainty that uranium gives off radiation?** (No. Lead students to reason that many more trials must be performed before a relatively certain conclusion can be established.)

of fluorescent uranium salt on top of the film and the paper and set both in the sun. Becquerel reasoned that if X-rays were produced by the uranium ② salt when it fluoresced, the X-rays would pass through the lightproof paper and produce an image on the film. The lightproof paper would prevent light from reaching the film and creating an image, however.

When Becquerel developed the film, he was delighted to see an image. The image was evidence that fluorescent substances give off X-rays when exposed to sunlight. In order to confirm his results, he prepared another sample of uranium salt and film to repeat his experiment the following day. But much to his disappointment, the next two days were cloudy. Impatient to get on with his work, Becquerel decided to develop the film anyway. What he saw on the film amazed him. Once again there was an image of the sample, even though the uranium salt had not been made to fluoresce. In fact, the image on the film was just as strong and clear as the image that had been formed when the sample was exposed to sunlight.

Becquerel realized that an invisible "something" given off by the salt had gone through the lightproof paper and produced an image. In time, this invisible "something" was named **nuclear radiation.** Becquerel tested many more uranium compounds and concluded that the source of nuclear radiation was the element uranium. An element that gives off nuclear radiation is said to be **radioactive.**

Marie Curie, a Polish scientist working in France and a former student of Becquerel's, became interested in Becquerel's pioneering work. She and her husband, French scientist Pierre Curie, began searching for other radioactive elements. In 1898, the Curies discovered a new radioactive element in a uranium ore known as pitchblende. They named the element polonium in honor of Marie Curie's native Poland. Later that year they discovered another radioactive element. They named this element radium, which means "shining element." Both polonium and radium are more radioactive than uranium. Since the Curies' discovery of polonium and radium, many other radioactive elements have been identified and even artificially produced.

Figure 5–2 *Marie Curie and her husband, Pierre, were responsible for the discovery of the radioactive elements radium and polonium. Since that time, many other radioactive elements have been identified.*

O ■ 113

ECOLOGY NOTE
FLUORESCENT SUBSTANCES

The French scientist Henri Becquerel suspected that fluorescent substances give off radiation. Have interested students research the "fluorescent" substances sometimes used to coat the hands of a watch or clock so that it is able to be seen in the dark. Are these substances really harmful, even in a minute way? Or are these substances not really fluorescent at all and harmless? Challenge students to find out more about these and other glow-in-the-dark substances and have them share their findings with the class.

der the exact same conditions? (Accept logical responses. Lead students to infer that it is now known that many radioactive materials give off harmful rays. Duplicating the conditions of Becquerel's experiment exactly would cause unnecessary exposure to radiation.)
• **Why do you think early experimenters did not wear protective clothing when working with radioactive materials?** (They were not aware of the potential danger of such material.)

As people years ago were unaware of some hazards, are people today also unaware of activities or materials that might prove hazardous in the future? Encourage students to speculate about the possibilities.

● ● ● ● **Integration** ● ● ● ●

Use the names given to newly discovered elements to integrate language arts concepts into your lesson.

● ● ● ● **Integration** ● ● ● ●

Use the introduction of the French scientist Henri Becquerel to integrate social studies concepts into your lesson.

Use the discovery of radioactivity to integrate concepts of electromagnetic radiation into your lesson.

ENRICHMENT

Have interested students perform library research about other major scientific discoveries involving radiation. Challenge these students to prepare a report that shows the role that various factors played in the discoveries as well as examples of evidence of the scientific method in each discovery. Findings should be shared with the class.

CONTENT DEVELOPMENT

Point out that science has learned many things about radiation since the time of its discovery. Ask:
• **Why shouldn't someone today attempt to duplicate Becquerel's experiment un-**

GAMMA RADIATION

When compared to the energy emitted by a hydrogen atom, the energy of a gamma radiation photon is staggering. Gamma photons carry more than 1 million times more energy than can be emitted by a typical hydrogen atom.

The Nature of Nuclear Radiation

Nuclear radiation cannot be seen. So radioactive elements were difficult to identify at first. But it was quickly realized that radioactive elements have certain characteristic properties. The first of these is the property observed by Becquerel. Nuclear radiation given off by radioactive elements alters photographic film. Another property of many radioactive elements is that they produce fluorescence in certain compounds. A third characteristic is that electric charge can be detected in the air surrounding radioactive elements. Finally, nuclear radiation damages cells in most organisms.

Today, scientists use the term **radioactivity** to describe the phenomenon discovered by Becquerel. **Radioactivity is the release of nuclear radiation in the form of particles and rays from a radioactive element.** The radiation given off by radioactive elements consists of three different particles or rays. The three types of radiation have been named alpha (AL-fuh) particles, beta (BAYT-uh) particles, and gamma (GAM-uh) rays after the first three letters of the Greek alphabet.

Figure 5–3 *The three types of radiation can be separated according to charge and penetrating power. When passed through a magnetic field, alpha particles are deflected toward the negative magnetic pole, beta particles are deflected toward the positive pole, and gamma rays are not deflected. Which type of radiation is the most penetrating?* ❶

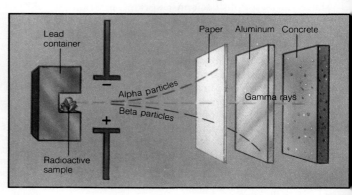

5–1 (continued)

CONTENT DEVELOPMENT

Remind students that the work by Becquerel was a spark used by others to discover more about the nature of radiation. Marie and Pierre Curie studied the phenomenon Becquerel discovered and named it radioactivity. Point out that Becquerel and the Curies discovered materials that were called natural radioactive elements—they emitted energy without adding energy to them. The work of these early researchers led others to conclude that these radioactive emissions consisted of three distinct types of radiation—alpha, beta, and gamma.

Stress that these three different kinds of radiation are distinguished according to how difficult they are to stop and by how they behave when subjected to electric and magnetic forces.

Explain that alpha particles are the least powerful (able to be stopped by a thin sheet of paper) and are positively charged. Beta particles are negatively charged and can be stopped by a sheet of aluminum several millimeters thick. Because of the opposite electrical charges of alpha and beta particles, they move in opposite directions when subjected to electric and magnetic influences. Gamma radiation is extremely penetrating and powerful but carries no charge because it is a form of electromagnetic energy.

● ● ● ● **Integration** ● ● ● ●

Use the relationship between the Greek alphabet and different types of radiation to integrate language arts concepts into your lesson.

Use the description of an alpha particle to integrate concepts of atomic structure into your lesson.

ALPHA PARTICLES An **alpha particle** is actually e nucleus of a helium atom—2 protons and 2 utrons. An alpha particle has a positive charge cause it contains 2 positive protons and no other arges. Alpha particles are the weakest type of nu-ear radiation. Although they can burn flesh, alpha rticles can be stopped by a sheet of paper.

BETA PARTICLES A **beta particle** is an electron. owever, a beta particle should not be confused th an electron that surrounds the nucleus of an m. A beta particle is an electron that is formed side the nucleus when a neutron breaks apart. ta particles have a penetrating ability 100 times eater than alpha particles. Beta particles can pass rough as much as 3 millimeters of aluminum.

GAMMA RAYS A **gamma ray** is an electromagnetic ve of extremely high frequency and short wave-ngth. Gamma rays are the same kind of waves as e visible light that enables you to see. That is, th are forms of electromagnetic waves. Gamma ys, however, carry a lot more energy. They are the st penetrating radiation given off by radioactive ements. Gamma rays can pass through several ntimeters of lead!

TYPES OF NUCLEAR RADIATION

Type	Atomic Mass	Atomic Number
Alpha (α)	4	2
Beta (β)	0	−1
Gamma (γ)	0	none

-1 Section Review

. Describe radioactivity. How did Becquerel discover radioactivity?
. How did the Curies use Becquerel's discovery?
. What is a radioactive element?
. Describe an alpha particle, a beta particle, and a gamma ray. How are they alike? How are they different?

ritical Thinking—*Forming a Hypothesis*

. There are several theories that attempt to ex-plain how a beta particle is produced. Develop a hypothesis to explain what a neutron—since it is neutral—may actually be composed of while still containing an electron.

O ■ 115

Answers

1 Gamma rays. (Interpreting diagrams)
2 The nucleus of a helium atom; an elec-tron formed inside a nucleus when a nu-cleus breaks apart; an electromagnetic wave of extremely high frequency and short wavelength. (Relating concepts)

Integration

1 Language Arts
2 Physical Science: Atomic Structure. See *Matter: Building Block of the Universe,* Chapter 4.

3. A radioactive element emits radiation consisting of alpha particles, beta parti-cles, and gamma rays.
4. An alpha particle is the nucleus of a he-lium atom, a beta particle is an electron formed inside a nucleus when a nucleus breaks apart, and a gamma ray is an elec-tromagnetic wave of extremely high fre-quency and short wavelength; they are similar in that all are types of nuclear ra-diation; they are different in their elec-trical charges and strengths.
5. Hypotheses will vary. Students might hypothesize that a neutron contains a pos-itive charge in addition to its electron, combining to yield an electrically neutral charge.

REINFORCEMENT/RETEACHING

Monitor students' responses to the Sec-tion Review questions. If students appear to have difficulty with any of the ques-tions, review the appropriate material in the section.

CLOSURE

▶ *Review and Reinforcement Guide*

At this point have students complete Section 5-1 in the *Review and Reinforcement Guide.*

REINFORCEMENT/RETEACHING

▶ *Activity Book*

Students who need practice on the con-ept of the behavior of radioactive particles nd rays should be provided with the Chap-er 5 activity called Alpha, Beta, and Gam-a Radiation in a Magnetic Field. In this ctivity students will explore the influence f magnetism on nuclear radiation.

INDEPENDENT PRACTICE

Section Review 5-1

1. Radioactivity is the release of matter and energy that results from changes in the nucleus of an atom; Becquerel ob-served that something given off by a ura-nium compound penetrated lightproof paper and produced an image on photo-graphic film in the absence of sunlight.
2. The Curies discovered two more ra-dioactive elements in the uranium ore known as pitchblende.

5-2 Nuclear Reactions

Have students investigate the percentage of power that your community receives from nuclear power plants. In various parts of the United States, nuclear power provides different percentages of the local power needs. Ask students whether they would like to have nuclear power plants near their home.

ESL STRATEGY 5-2

Have students write a paragraph explaining how radioactivity occurs and why Becquerel and the Curies had difficulty understanding the origin of this radioactivity.

After using their dictionaries to find the different meanings of *binding* and *stable*, have students perform these activities:
1. Explain why binding energy is an appropriate chemical term.
2. Write two sentences using different meanings of *stable*.
3. Explain why atoms with unstable nuclei will come apart.

Guide for Reading

Focus on these questions as you read.
▶ *Why do radioactive nuclei undergo nuclear reactions?*
▶ *How are radioactive decay and artificial transmutation similar and how are they different?*

Helium nucleus

Lithium nucleus

Carbon nucleus

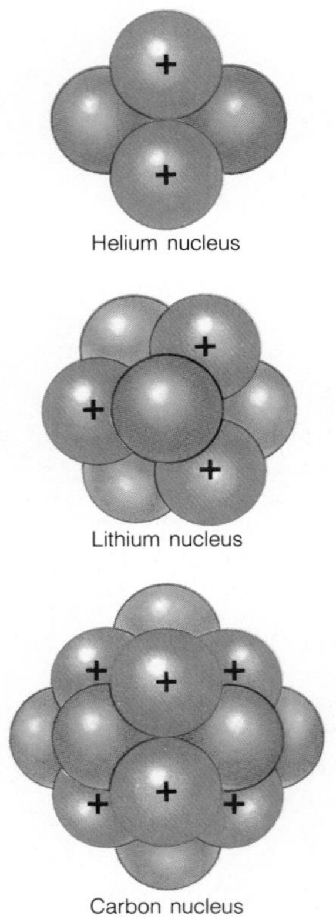

Figure 5-5 *The nucleus of an atom contains positively charged protons and neutral neutrons that are held together by the nuclear strong force. A helium nucleus has 2 protons and 2 neutrons. How many protons and neutrons does a lithium nucleus contain? A carbon nucleus?* ❶

116 ■ O

5-2 Nuclear Reactions

Although Becquerel and the Curies observed radioactivity, they could not explain its origin. The reason for this is understandable: The source of radioactivity is the nucleus of an atom. But Becquerel discovered radioactivity well before the nucleus was discovered. Several years after Becquerel's and the Curies' work, it was determined that radioactivity results when the nuclei of atoms of certain elements change, emitting particles and/or rays. What still remained unknown, however, was what makes a nucleus break apart and why only some elements are radioactive.

Nuclear Stability

The answers to these puzzling questions would be found in the atom—specifically, in the nucleus. The nucleus of an atom contains protons and neutrons. Protons are positively charged particles. Neutrons are neutral particles; they have no charge. It is a scientific fact that particles with the same charge (positive or negative) repel each other. Thus protons repel each other. How, then, does the nucleus hold together? A force known as the **nuclear strong force** overcomes the force of repulsion between protons and holds protons and neutrons together in the nucleus. The energy associated with the strong force is called **binding energy.**

The binding energy is essential to the stability of a nucleus. In some atoms, the binding energy is great enough to hold the nucleus together permanently. The nuclei of such atoms are said to be stable. In other atoms, the binding energy is not as great. The nuclei of these atoms are said to be unstable. An unstable nucleus will come apart. Atoms with unstable nuclei are radioactive.

Some elements that are not radioactive have radioactive forms, or **isotopes** (IGH-suh-tohps). What

---

TEACHING STRATEGY 5-2

FOCUS/MOTIVATION

Ask students to name examples of dramatic changes that matter can undergo, such as burning. Then point out that in almost all such changes, there is simply a rearrangement of atoms but no change in the identity of atoms. Point out that it is possible for atoms themselves to change in kind and that such change, called transmutation, is rarely observed under ordinary conditions.

CONTENT DEVELOPMENT

Review the nature of subatomic particles.
• **What is the charge and mass of a proton?** (A proton is a positively charged particle; it has a mass of 1 amu.)
• **What is the charge and mass of a neutron?** (A neutron is an electrically neutral particle; it has a mass of 1 amu.)

Also review the concepts of atomic number and mass number.
• **What is the atomic number of an element?** (It is the number of protons in the nucleus of an atom.)
• **What is the mass number of an element?** (It is the sum of the protons and the neutrons in the nucleus of an atom.)

Make certain that students understand the concept of isotopes as atoms that have the same number of protons but different numbers of neutrons.

• **Do particles with like charges attract or repel each other?** (Repel.)
• **The protons in the nucleus of an atom have positive charges. How can a nucleus of an atom hold together if all the protons are repelling one another?** (A force called the nuclear strong force holds the nucleus together; the energy associated with this nuclear strong force is called binding energy.)

NONRADIOACTIVE AND RADIOACTIVE ISOTOPES OF SOME COMMON ELEMENTS

Element	Nonradioactive Isotope	Radioactive Isotope
Hydrogen	1 proton 0 neutrons	1 proton 2 neutrons
Helium	2 protons 2 neutrons	2 protons 4 neutrons
Lithium	3 protons 4 neutrons	3 protons 5 neutrons
Carbon	6 protons 6 neutrons	6 protons 8 neutrons
Nitrogen	7 protons 7 neutrons	7 protons 9 neutrons
Oxygen	8 protons 8 neutrons	8 protons 6 neutrons
Potassium	19 protons 20 neutrons	19 protons 21 neutrons

Figure 5–6 *An isotope is an atom of an element that has the same number of protons but a different number of neutrons. Often, if the number of neutrons greatly differs from the number of protons, the isotope does not have enough binding energy to hold the nucleus together and is therefore radioactive.*

isotope? The number of protons in the atoms of a particular element cannot vary. An atom is identified by the number of protons it contains. (The number of protons is called the atomic number.) Carbon atoms would not be carbon atoms if they had 5 protons or 7 protons—only 6 protons will. Yet there are some carbon atoms that have 6 neutrons, and others that have 8 neutrons. The difference in the number of neutrons affects the characteristics of the atom but not its identity. Atoms that have the same number of protons (atomic number) but different numbers of neutrons are called isotopes.

Many elements have at least one radioactive isotope. For example, carbon has two common isotopes—carbon-12 and carbon-14. Carbon-12, which you are familiar with as coal, graphite, and diamond, is not radioactive. Carbon-14, used in dating fossils, is radioactive. Figure 5–6 shows the radioactive and nonradioactive isotopes of some common elements.

ACTIVITY
WRITING

What Is a Quark?

Scientists have proposed that all nuclear subatomic particles are composed of basic particles called quarks. Using books and other reference materials in the library, look up the word quark. Find out what quarks are and how they were discovered. Describe some different types of quarks and what they do. Find out about the recent discoveries that have been made involving quarks. Report your findings to the class.

ACTIVITY
WRITING
WHAT IS A QUARK?

Skills: Relating facts, applying concepts

This is a challenging activity that will allow students to investigate some recent concepts of particle physics. Because much of the information concerning quarks is extremely sophisticated, encourage students to present highlights of their findings, such as the fact that protons and neutrons are thought to be made up of combinations of three quarks. You may wish to point out to students that "free" quarks have never been observed and that current theory suggests that quarks cannot exist singly.

Integration: Use this Activity to integrate language arts skills into your science lesson.

What happens if there is not enough binding force in an isotope to hold a nucleus together? (The isotope becomes radioactive.)

● ● ● ● **Integration** ● ● ● ●

Use the description of carbon isotopes to integrate concepts of radioactive dating into your lesson.

GUIDED PRACTICE

Skills Development
Skill: Making comparisons

Ask interested students to construct a graph of average mass (Y axis) versus atomic number (X axis) for the elements or selected elements. The required data for this activity can be obtained from any chart of the elements or periodic table. Point out that the atomic mass of an element is, numerically speaking, roughly equivalent to a weighted average, by occurrence, of the mass numbers of the isotopes of that element. After their work is completed, ask students to interpret their graphs, which should clearly illustrate that the ratio of atomic mass to atomic number (and therefore the appropriate ratios of mass number to atomic number and of number of neutrons to number of protons) tends to increase as atomic mass increases.

ACTIVITY

The Nuclear Age

Often when you learn about discoveries or developments in science, you do not consider the lives and personalities of the scientists involved or the conditions of the time period during which they worked. These factors are important in appreciating the significance of the discovery. The production of the atomic bomb at the end of World War II is a good example of a discovery that grew out of the expectations and concerns of the people involved.

After completing thorough research, write a paper describing the events that led to an enormous research project known as the Manhattan Project. Be sure to include information about these people: Enrico Fermi, Otto Hahn, Lise Meitner, Otto Frisch, Albert Einstein, and President Franklin D. Roosevelt. Also discuss the impact of these developments on history.

Becoming Stable

Imagine a large rock hanging over the edge of a cliff. How might you describe the rock's precarious position? You would probably say it is unstable, meaning that it cannot remain that way for long. Most likely, the rock will fall to the ground below, where it will be in a stable condition. Once it has fallen, the rock will certainly never move itself back to the cliff. Perhaps now you can think of an answer as to why an unstable nucleus breaks apart.

A nucleus that is unstable can become stable by undergoing a nuclear reaction, or change. There are four types of nuclear reactions that can occur. In each type, the identity of the original element is changed as a result of the reaction. You will now learn about each type of nuclear reaction.

Radioactive Decay

The process in which atomic nuclei emit particles or rays to become lighter and more stable is called **radioactive decay.** Radioactive decay is the spontaneous breakdown of an unstable atomic nucleus. There are three types of radioactive decay, each determined by the type of radiation released from the unstable nucleus.

ALPHA DECAY Alpha decay occurs when a nucleus releases an alpha particle. The release of an alpha particle (2 protons and 2 neutrons) decreases the mass number of the nucleus by 4. The mass number is the sum of the number of protons and neutrons in the nucleus. Each proton and each neutron has a mass of 1. The release of an alpha particle decreases the number of protons, or the atomic number, by . Thus the original atom is no longer the same. A new atom with an atomic number that is 2 less than the original is formed.

An example of an element that undergoes alpha decay is an isotope of uranium called uranium-238. The number 238 to the right of the hyphen is the mass number for this particular nucleus. An isotope of an element is often represented by using the element's symbol, mass number, and atomic number. The mass number is written to the upper left of the symbol. At the lower left, the atomic number (or

---

5–2 (continued)

CONTENT DEVELOPMENT

To help students better understand nuclear changes, have them first recall physical and chemical changes by asking:

• **What is a physical change?** (A physical change is a change in phase. Point out that these physical changes in matter typically include melting, freezing, vaporization, and condensation.)

• **What are several examples of physical changes?** (Examples will vary. Students might suggest changing ice into water or water into ice.)

• **What is a chemical change?** (A chemical change is the process by which a substance changes.)

• **What are several examples of chemical changes?** (Examples will vary. Students might suggest burning a match or burning a piece of paper.)

Point out that in chemical changes, the particles of one substance become rearranged to form a new substance, but the same number of particles and atoms exists before and after the reaction.

Explain that in a nuclear change, the nucleus of an atom changes so that a new element is formed. Emphasize that this process is called transmutation. Remind students that transmutation is rarely observed under ordinary conditions.

INDEPENDENT PRACTICE

▶ *Activity Book*

Students who need practice on the concept of transmutation should complete the chapter activity Decay Series of Uranium-238. In this activity students will determine the element or number missing from various positions in a U-238 decay series.

number of protons) is written. Uranium has 92 protons. So this is the way uranium-238 would be represented:

$$^{238}_{92}\text{U}$$

The number of neutrons in the nucleus can be determined by subtracting the number of protons from the mass number. In this example, the number of neutrons is the mass number 238 minus the number of protons, 92, or 146 (238 − 92 = 146).

When uranium-238 undergoes alpha decay, or loses an alpha particle, it changes into an atom of thorium (Th), which has 90 protons and 144 neutrons. What is the mass number of thorium? ❶

BETA DECAY Beta decay occurs when a beta particle is released from a nucleus. As you have learned, a beta particle is an electron formed inside the nucleus when a neutron breaks apart. The other particle that forms when a neutron breaks apart is a proton. So beta decay produces a new atom with the same mass number as the original atom but with an atomic number one higher than the original atom. The atomic number is one higher because there is now an additional proton.

An example of an element that undergoes beta decay is carbon-14. An atom of carbon-14 has 6 protons and 8 neutrons. During beta decay it changes into an atom of nitrogen-14. An atom of nitrogen-14 has 7 protons and 7 neutrons.

When a nucleus releases either an alpha particle or a beta particle, the atomic number, and thus the identity, of the atom changes. **The process in which one element is changed into another as a result of changes in the nucleus is known as transmutation.** The word **transmutation** comes from the word *mutation*, which means change, and the prefix *trans-*, which means through.

GAMMA DECAY Alpha and beta decay are almost always accompanied by gamma decay, which involves the release of a gamma ray. When a gamma ray is emitted by a nucleus, the nucleus does not change into a different nucleus. But because a gamma ray is an extremely high-energy wave, the nucleus makes a transition to a lower energy state.

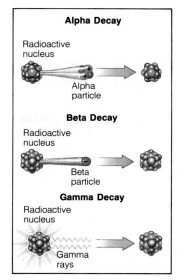

Figure 5-7 *Radioactive elements emit mass and energy during three different types of decay processes. During alpha decay, a helium nucleus and energy are released. A beta particle, or electron, and energy are released during beta decay. During gamma decay, high-energy electromagnetic waves are emitted. Which type of decay does not result in a different element?* ❷

O ■ 119

BACKGROUND INFORMATION

NATURE'S FUNDAMENTAL FORCES

The strong nuclear force is one of the four so-called fundamental forces in nature. The other three forces are the weak nuclear force, which accounts for phenomena such as beta decay of unstable nuclei, electromagnetic force, and gravitational force. It has been hypothesized that the strong force is mediated, or carried, by a particle called the gluon. Gluons are believed to hold together quarks, which are hypothesized to make up most subatomic particles.

Remind students that during beta decay, an element with the same mass number as the original atom forms, but with an atomic number of one more than before the decay.
• **Why would the element formed during beta decay have the same mass number but a greater atomic number?** (Mass number remains the same because even though a neutron is lost, a proton is gained. The atomic number increases due to the additional proton.)

Emphasize that once a nucleus decays and releases either an alpha particle or a beta particle, the nucleus becomes stable. This changing of one element into another is known as transmutation.

● ● ● ● **Integration** ● ● ● ●

Use the explanation of the word *transmutation* to integrate language arts concepts into your science lesson.

CONTENT DEVELOPMENT

Point out that radioactive decay releases both particles and energy. The particles and the energy that are released are both referred to as nuclear radiation. Explain that the specific way that particles and energy are released from the unstable nucleus of a radioactive element determines the type of decay it experiences. Point out that there are three kinds of radiation that can be released during decay—alpha particles, beta particles, and gamma rays.

After students have had an opportunity to read about the various types of decay, ask:
• **What charge does an alpha particle have?** (Positive.)
• **What can be said about an atom after the release of an alpha particle?** (A new atom is formed with an atomic number that is two less than the original atom.)

ISOTOPES

Isotopes can be represented in two ways. In the first way, the mass number is written after the name of the element: for example, carbon-12, uranium-238. In the second way, the mass number is written as a superscript at the upper left of the atomic symbol, and the atomic number is written as a subscript at the lower left of the atomic symbol.

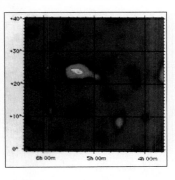

Figure 5–8 *Gamma rays are not as rare as you may think. For example, energy in the form of gamma rays is constantly being emitted from objects in space. This image of a solar flare was recorded by detecting gamma rays from the sun.*

Radioactive Half-Life

A sample of any radioactive element consists of a vast number of radioactive nuclei. These nuclei do not all decay at one time. Rather, they decay one by one over a period of time at a fixed rate. The fixed rate of decay of a radioactive element is called the **half-life.** The half-life is the amount of time it takes for half the atoms in a given sample of an element to decay.

The half-life of carbon-14 is 5730 years. In 5730 years, half the atoms in a given sample of carbon-14 will have decayed to another element: nitrogen-14. In yet another 5730 years, half the remaining carbon-14 will have decayed. At that time, one fourth—or one half of one half—of the original sample will be left. One fourth of the original sample will be carbon-14, and three fourths will be nitrogen-14.

Suppose you had 20 grams of pure barium-139. Its half-life is 86 minutes. So after 86 minutes, half the atoms in the sample would have decayed into another element: lanthanum-139. You would have 10 grams of barium-139 and 10 grams of lanthanum-139. After another 86 minutes, half the atoms in the 10 grams of barium-139 would have decayed into lanthanum-139. You would then have 5 grams of barium-139 and 15 grams of lanthanum-139. What would you have after the next half-life? ❶

Figure 5–9 *The half-life of a radioactive element is the amount of time it takes for half the atoms in a given sample of the element to decay. After the first half-life, half the atoms in the sample are the radioactive element. The other half are the decay element, or the element into which the radioactive element changes. What remains after the second half-life? After the third?* ❷

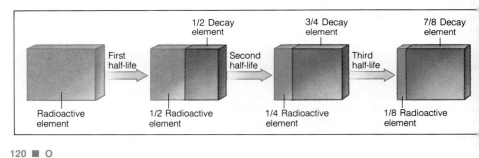

5–2 (continued)

CONTENT DEVELOPMENT

Explain that radioactive elements decay at differing rates and that the amount of time it takes for one half of the atoms in a radioactive element to decay is called the half-life.

● ● ● ● **Integration** ● ● ● ●

Use the example of the radioactive decay of barium-139 to integrate mathematics concepts into your science lesson.

GUIDED PRACTICE
Skills Development
Skill: Applying concepts

Show an empty 250-mL beaker to students. Fill it with a colored solution, such as colored water, and tell students to imagine that the beaker contains an element with an unstable nucleus. Have a volunteer use a stopwatch to measure the time it takes you to slowly pour one half (125 mL) of the solution into a separate beaker. Then show students the original beaker with one half of the solution remaining in it.

• **How much of the original element remains?** (One half.)

• **How much time did it take to reduce the original amount by one half?** (Accept the correct answer.)

Repeat the demonstration using a 500-mL beaker and a solution of a different color. Pour faster so that the amount of time is less than in the first demonstration and point out that it took a different amount of time to reduce the solution by one half.

• **Which decay model had the shortest half-life?** (The second model.)

• **What is a half-life?** (It is the amount of time it takes for half of the starting atoms to decay into other elements.)

Based on the data gathered from these models, have students determine how much original material would remain af-

The half-lives of certain radioactive isotopes are useful in determining the ages of rocks and fossils. Scientists can use the half-life of carbon-14 to determine the approximate age of organisms and objects less than 50,000 years old. The technique is called carbon-14 dating. Other radioactive elements, such as uranium-238, can be used to date objects many millions of years old.

Half-lives vary greatly from element to element. Some half-lives are only seconds; others are billions of years. For example, the half-life of rhodium-106 is 30 seconds. The half-life of uranium-238 is 4.5 billion years!

Decay Series

As radioactive elements decay, they change into other elements. These elements may in turn decay, forming still other elements. The spontaneous breakdown continues until a stable, nonradioactive nucleus is formed. The series of steps by which a radioactive nucleus decays into a nonradioactive nucleus is called a **decay series.** Figure 5–11 on page 122 shows the decay series for uranium. What stable nucleus results from this decay series? ❸

Because of the occurrence of decay series, certain radioactive elements are found in nature that otherwise would not be. In the 5-billion-year history of the solar system, many isotopes with short half-lives have decayed quickly. Thus they should not exist in

HALF-LIVES OF SOME RADIOACTIVE ELEMENTS

Element	Half-Life
Bismuth-212	60.5 minutes
Carbon-14	5730 years
Chlorine-36	400,000 years
Cobalt-60	5.26 years
Iodine-131	8.07 days
Phosphorus-32	14.3 days
Polonium-215	0.0018 second
Polonium-216	0.16 second
Radium-226	1600 years
Sodium-24	15 hours
Uranium-235	710 million years
Uranium-238	4.5 billion years

Figure 5–10 *The half-lives of radioactive elements vary greatly. Using the known half-lives of certain radioactive elements, such as carbon-14 and uranium-238, scientists can determine the age of ancient objects. Radioactive dating has been essential to the discovery of information about the Earth's history and about the evolution of organisms such as this ancient turtle.*

O ■ 121

● ● ● ● **Integration** ● ● ● ●

Use the process of radioactive dating to integrate evolution concepts into your lesson.

GUIDED PRACTICE

Skills Development

Skills: Applying concepts, relating facts

At this point have students complete the in-text Chapter 5 Laboratory Investigation: The Half-Life of a Sugar Cube. In this investigation students will determine the half-life of a large sample of sugar cubes.

REINFORCEMENT/RETEACHING

▶ *Activity Book*

Students who need practice on the concept of radioactive elements should be provided with the Chapter 5 activity called Radioactive and Nonradioactive Elements at a Glance. In this activity students will create a copy of the periodic table and will use different colors to highlight the elements that are radioactive, nonradioactive, and synthetic.

er, for example, the third and fifth half-lives of the decay series.

CONTENT DEVELOPMENT

Remind students that during radioactive decay, particles and energy are released from an unstable atomic nucleus and that radioactive decay also creates lighter and more stable nuclei. Also have students recall that alpha and beta decay are almost always accompanied by gamma decay, which involves the release of a gamma ray.

Explain that one of the useful tools of decay is dating, or determining the approximate age of, fossils and other objects. Point out that the radioactive isotope carbon-14 is used to determine, or date, the age of "young" fossils and objects—those less than 50,000 years old—and that the radioactive isotope uranium-238 is used to determine, or date, the age of "old" fossils or objects—those more than 50,000 years old. Emphasize that this radioactive dating process determines an approximate age only.

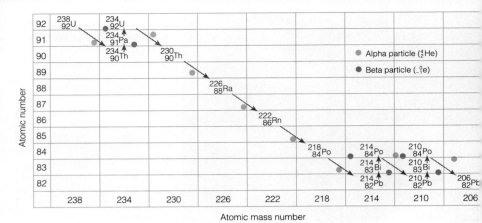

HISTORICAL NOTE

ALCHEMISTS

A goal of alchemists—early philosopher-scientists—was to transmutate, or change, so-called base metals into gold. Alchemists had no knowledge of the nucleus or its changes. As a result, alchemists could produce only physical and chemical rather than nuclear changes.

Figure 5–11 *The decay series for uranium-238 is shown in this graph. Radioactive uranium forms nonradioactive lead as a result of the decay series. What happens to the atomic mass number during a decay series? The atomic number?* ❶

Activity Bank

The Domino Effect, p.159

nature today. This is hardly the case, however. For example, radium, whose half-life is 1600 years, should have disappeared long ago. Yet it still exists on Earth today. This is because radium is part of t decay series for an isotope with a much longer hal life: uranium-238. Recall that uranium-238 has a ha life of 4.5 billion years.

Artificial Transmutation

Once scientists understood how natural transm tation occurred, they worked to produce **artificial transmutation.** The key was to find a way to chang the number of protons in the nucleus of an atom. Ernest Rutherford, the same scientist who discover the nucleus of the atom, produced the first artifici transmutation. By using alpha particles emitted du ing the radioactive decay of radium to bombard (h forcefully) nitrogen nuclei, he produced an isotop of oxygen.

Getting the particles to hit the target nuclei wi enough force to alter them is extremely difficult. I order to more effectively bombard nuclei with high energy particles, scientists have developed devices accelerating (speeding up) charged particles. One such device is the supercollider you read about at the beginning of this chapter. Other devices are th cyclotron, synchrotron, betatron, and linear accele tor. These devices use magnets and electric fields speed up particles and produce collisions.

5–2 (continued)

ENRICHMENT

▶ *Activity Book*

Students will be challenged by the Chapter 5 activity in the *Activity Book* called Atomic Arithmetic. In this activity students will use facts about radioactive elements and computations to determine the half-life of carbon-14.

GUIDED PRACTICE

▶ *Laboratory Manual*

Skills Development

Skill: Applying concepts

At this point you may want to have students complete the Chapter 5 Laboratory Investigation in the *Laboratory Manual* called Half-life of a Capacitor. In this investigation students will determine how the half-life of a discharging capacitor can be determined.

ENRICHMENT

Students might be interested to lear that there is no known way to predict ex actly when a given atom in a radioactiv sample will decay. Half-life, which dea with decay-time considerations such a this, is a statistical concept, and its prol ability can be applied only to large num bers of atoms in a sample. For exampl if the half-life of an isotope is one hou all that can be said is that one half of th atoms in a sample of that isotope will hav

Before the discovery of the neutron in 1932, mainly alpha particles and protons were used as the "bullets" to bombard nuclei. But because both these particles are positively charged, they are repelled by the positive charge of the target nucleus. A large amount of extra energy is required simply to overcome this repulsion.

Enrico Fermi (FER-mee), an Italian scientist, and his co-workers realized that because neutrons are neutral, they are not repelled by the nucleus. These researchers discovered that neutrons can penetrate the nucleus of an atom more easily than a charged particle can. Neutrons can go through the nucleus without changing it; they can cause the nucleus to disintegrate; or they can become trapped by the nucleus, causing it to become unstable and break apart.

After a great deal of experimentation, the elements neptunium and plutonium were created. They were the first **transuranium elements.** Transuranium elements (also known as synthetic elements) are those with more than 92 protons in their nuclei. In other words, transuranium elements have atomic numbers greater than 92. A whole series of transuranium elements have been formed by bombarding atomic nuclei with neutrons, alpha particles, or other nuclear "bullets."

Figure 5–12 *Artificial transmutation of elements is done in a particle accelerator, such as the one at Fermilab in Illinois. This aerial view shows the outline of the underground tunnel. Particles traveling through long tubes will reach a final speed greater than 99.999 percent of the speed of light!*

O ■ 123

decayed by the end of an hour, but it cannot be predicted which of the atoms will have decayed.

CONTENT DEVELOPMENT

Point out that spontaneous decay, or transmutation, occurs naturally in unstable atoms. The transmutation of stable atoms, however, can be accomplished by applying an outside force. The process of using machines and laboratory procedures to transmute elements is known as artificial transmutation.

Artificial transmutation can be accomplished in a variety of ways, although a tremendous amount of energy is required in all cases. Early work in the field of artificial transmutation involved using high-speed charged particles as "bullets" to bombard nuclei. These positively charged or negatively charged particles are accelerated at speeds approaching the speed of light, then fired into various materials to be examined.

Point out that the Italian scientist Enrico Fermi used the existing field of knowledge at the time to help develop a particle accelerator that used neutrons as "bullets." Because neutrons have a neutral electrical charge, they were influenced less by existing charges in an atom, and they became a successful evolution of the particle accelerator.

Emphasize to students that the early accelerator was quite small and simple when compared with the artificial transmutation designs of today. Have interested students research the process of artificial transmutation further by gathering information about cyclotrons, synchrotrons, betatrons, or linear accelerators, and by reporting their findings to the class.

ACTIVITY
THINKING
INTERPRETING NUCLEAR EQUATIONS

Skill: Relating concepts

This activity will help students explore equations showing an artificial transmutation and selected portions of the decay series for uranium-238. Students should infer from the equations that a uranium-235 atom in conjunction with the bombardment of a neutron yields rubidium, cesium, and 2 neutrons; lead-210 decays to bismuth and a beta particle; and radon decays to polonium and an alpha particle.

5–2 (continued)

ENRICHMENT

▶ *Activity Book*

Students will be challenged by the Chapter 5 activity in the *Activity Book* called Nuclear Equations. In this activity students will complete various nuclear equations by filling in blank spaces with the correct information.

CONTENT DEVELOPMENT

Point out that radioactive isotopes of natural elements are created by bombarding the nucleus of stable atoms with neutrons. An example of a radioactive isotope with a practical application is cobalt-60. Cobalt-60 emits beta particles and gamma rays during the course of its decay, and this energy, or radiation, is used to treat tumors by radiation therapy. In some cases, low-level dosages of cobalt-60 are sometimes physically implanted in tumors.

ACTIVITY

Interpreting Nuclear Equations

Nuclear reactions can be described by equations in much the same way chemical reactions can. Symbols are used to represent atoms as well as particles. And the total mass numbers on both sides of the equation are equal. The following reaction describes the alpha decay of uranium-238.

$$^{238}_{92}\text{U} \rightarrow \,^{234}_{90}\text{Th} + \,^{4}_{2}\text{He} + \text{Gamma rays}$$

Notice how an alpha particle is written as a helium nucleus. Other particles are written in a similar manner: a beta particle is written $^{0}_{-1}\text{e}$, and a neutron is written $^{1}_{0}\text{n}$.

Describe what is happening in each of the following:

$$^{235}_{92}\text{U} + \,^{1}_{0}\text{n} \rightarrow \,^{90}_{37}\text{Rb} + \,^{144}_{55}\text{Cs} + 2\,^{1}_{0}\text{n}$$

$$^{210}_{82}\text{Pb} \rightarrow \,^{210}_{83}\text{Bi} + \,^{0}_{-1}\text{e}$$

$$^{222}_{86}\text{Rn} \rightarrow \,^{4}_{2}\text{He} + \,^{218}_{84}\text{Po}$$

124 ■ O

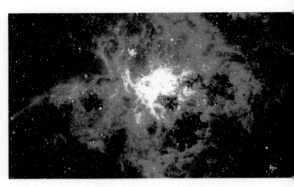

Figure 5–13 *Some elements present at the origin of the universe no longer exist. Others, whose half-lives are so short that they should no longer exist, do exist because they are part of the decay series of other radioactive elements. Scientists study decay series in an effort to learn more about the universe and its structures, such as the Tarantula Nebula.*

Radioactive isotopes of natural elements can be made by using a similar technique. Marie Curie's daughter, Irene, and Irene's husband, Frederic Joliot, discovered that stable atoms can be made radioactive when they are bombarded with neutrons. For example, by shooting neutrons at the nucleus of an iodine atom, scientists have been able to make I-131, a radioactive isotope of iodine.

5–2 Section Review

1. How is the binding energy related to the stability of a nucleus? How can an unstable nucleus become stable?
2. What happens during radioactive decay? What are three types of radioactive decay?
3. What is half-life? What is a decay series?
4. What is transmutation? Artificial transmutation?

Critical Thinking—*Making Calculations*
5. The half-life of radium-222 is 38 seconds. How many grams of radium-222 remain in a 12-gram sample after 76 seconds? After 114 seconds? How many half-lives have occurred when 0.75 gram remains?

● ● ● ● **Integration** ● ● ● ●

Use Figure 5–13 to integrate concepts of astronomy into your lesson.

REINFORCEMENT/RETEACHING

To reinforce the concept of radioactive half-life, have groups complete the following exercise. (For this activity, each group will require about 100 small objects of the same size, such as marbles, buttons, or pennies; a piece of posterboard; and tape or glue.) Instruct each group to

1. Illustrate half-life by letting each small object represent one atom. (Point out that they should choose the starting number of objects carefully—it will be difficult to divide the small objects into halves and other fractional parts in successive half-lives.)
2. Display the "atoms" on posterboard to show how the number of atoms is reduced in each half-life.
3. Decide on a half-life for each group's sample. Label each group of atoms with the time it has taken to reach that num-

CONNECTIONS

❷ An Invisible Threat

When you think of the greatest threats to the environment and to your health and safety, scenes of polluted waterways and dirty landfills probably come to mind. So you might be surprised to learn that one of the most widespread and serious environmental threats cannot even be seen. This dangerous menace is radon.

Radon-222 is a radioactive element (atomic number 86) that is produced as part of the decay series for uranium-238. Because of uranium's long half-life, radon is continuously being generated. Since the mid-1980s, radon has become a priority concern for the United States Environmental Protection Agency (EPA). In addition to producing the effects normally associated with nuclear radiation, highly concentrated radon in the air can cause lung cancer if inhaled in large quantities. Radon represents one of the few naturally existing pollutants.

The EPA has set limits for radon levels in the home. However, many homes have radon levels nearly 1000 times greater than the limit. One reason for this is that many buildings are constructed on land rich in uranium ore. Radon is released from soil and rocks containing this ore. In the outdoors, radon is diluted to safe levels. But when it leaks into buildings through cracks in basement floors and walls, radon is trapped—and dangerous.

The concentration of radon in a building depends on the type of construction

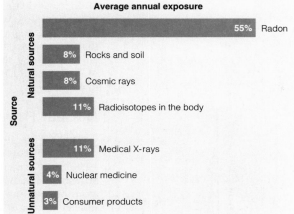

Average annual exposure

Source

Natural sources:
- 55% Radon
- 8% Rocks and soil
- 8% Cosmic rays
- 11% Radioisotopes in the body

Unnatural sources:
- 11% Medical X-rays
- 4% Nuclear medicine
- 3% Consumer products

and the materials used. New energy-efficient buildings, which are designed to keep heated or cooled air in, also trap radon. One way to protect against radon, especially in areas with a high natural concentration, is to seal cracks in foundation walls and floors. Another way is to improve air ventilation by circulating outside air into a building, thereby diluting existing radon concentrations.

Over the long term, everyone is exposed to potentially damaging radiation from both natural sources and human activity. But radon exposure currently accounts for more than half the average annual exposure to radiation. Home radon-testing kits have become a common household item in many parts of the country. One of the most important protections against the health hazards of radon is the awareness that something that cannot be seen, touched, or smelled can be life-threatening.

This feature allows students the opportunity to connect the concept of radioactivity to a real-life potential threat: radon gas. As the contents of this article are discussed, remind students that no one is immune from the threat of radon gas, yet its threat is significantly diminished with sensible actions. Point out that radon is a naturally occurring gas that is also emitted from unnatural sources. Whether radon is found naturally in the Earth or unnaturally, it becomes harmful when its concentration is allowed to increase within an enclosed space. After students have described actions that should be enlisted to minimize the threat of radon-gas exposure, point out that the reliability of some home-testing kits for radon gas has been questioned. Discuss different things a consumer might do to help select a reliable testing kit for radon-gas detection.

If you are teaching thematically, you may want to use the Connections feature to reinforce the themes of energy, patterns of change, scale and structure, systems and interactions, and stability.

Integration: Use the Connections feature to integrate ecology into your lesson.

ber, how many "atoms" remain, and which half-life is represented.

INDEPENDENT PRACTICE

Section Review 5–2

1. If the binding energy of a nucleus is great enough, the nucleus is stable and will be held together permanently; if the binding energy of a nucleus is not sufficient, the nucleus is unstable and will come apart. A nucleus that is unstable can become stable by undergoing a change or a nuclear reaction such as radioactive decay.

2. During radioactive decay, an unstable atomic nucleus spontaneously breaks down, causing the nucleus to emit particles or rays and become lighter and more stable. The three types of radioactive decay are alpha decay, beta decay, and gamma decay.

3. A half-life is the amount of time it takes for one half of the atoms in a sample of a radioactive element to decay. A decay series is the spontaneous breakdown of an unstable atomic nucleus until such time as

a stable, nonradioactive nucleus is formed.

4. Transmutation is the process by which the nucleus of an atom changes so that a new element is formed. Artificial transmutation is the process by which charged particles bombard an atomic nuclei, forcing it to absorb additional protons or neutrons, producing new elements.

5. 3 grams; 1.5 grams; four.

REINFORCEMENT/RETEACHING

Review students' responses to the Section Review questions. Reteach any material that is still unclear, based on students' responses.

CLOSURE

▶ *Review and Reinforcement Guide*

Students may now complete Section 5–2 in the *Review and Reinforcement Guide*.

5–3 Harnessing the Nucleus

Have students investigate the percentage of power that various nations of the world receive from nuclear power plants. Many countries of the world rely on a much higher percentage of nuclear-generated power than does the United States. You might ask your students to suggest why the US relies so much on fossil fuels.

ESL STRATEGY 5–3

Help students remember the difference between nuclear fission and nuclear fusion by explaining that the term *fission* means "a division or a splitting apart" and *fusion* means "a melting together." Then have students identify which of these terms is described

• when 2 atomic nuclei of smaller masses come together to form a single nucleus of larger mass

• when an atomic nucleus splits into 2 smaller nuclei of approximately equal size

Ask students to list three advantages that nuclear fusion has compared to nuclear fission.

TEACHING STRATEGY 5–3

FOCUS MOTIVATION

Ask students if they have ever heard of the term "chain letter" or if they have ever known someone who received a chain letter. Then ask:

• **What does a chain letter tell someone to do?** (Accept logical responses. Students might suggest that the letter will ask someone to send the same letter to someone else; to send something to the person at the top of the list; to do something and not break the chain; and so on.)

• **If you were to receive a letter and sent the same letter to ten different people and each of those ten people in turn sent the same letter to ten different people, how many times would the letter have been sent?** (If students have difficulty answering the question, encourage them to draw a diagram of it or have a volunteer

---

Guide for Reading

Focus on this question as you read.

▶ What is the difference between nuclear fission and nuclear fusion?

Figure 5–14 *In the midst of the social and political strife of the late 1930s, scientists such as Lise Meitner and Otto Hahn succeeded in recognizing and explaining the events of nuclear fission. What is nuclear fission?* **1**

126 ■ O

5–3 Harnessing the Nucleus

Radioactive decay and the bombardment of a nucleus with particles are two ways in which energy is released from the nucleus of an atom. The amount of energy released, however, is small compared with the tremendous amount of energy known to bind the nucleus together. Long ago, scientists realized that if somehow they could release more of the energy holding the nucleus together, huge amounts of energy could be gathered from tiny amounts of mass.

Nuclear Fission

During the 1930s, several other scientists built upon the discovery of Fermi and his co-workers. In 1938, the German scientists Otto Hahn and Fritz Strassman discovered that when the nucleus of an atom of uranium-235 is struck by a neutron, two smaller nuclei of roughly equal mass were produced. Two other scientists, Lise Meitner and Otto Frisch, provided an explanation for this event: The uranium nucleus had actually split into two. What made the discovery and the explanation so startling was that until then the known nuclear reactions had involved only knocking out a tiny fragment from the nucleus—not splitting it into two!

This reaction—the first of its kind ever to be produced—is an example of **nuclear fission** (FIHSH-uhn). It was so named because of its resemblance to cell division, or biological fission. **Nuclear fission is the splitting of an atomic nucleus into two smaller nuclei of approximately equal mass.** Unlike radioactive decay, nuclear fission does not occur spontaneously.

In one typical fission reaction, a uranium-235 nucleus is bombarded by a neutron, or nuclear "bullet." The products of the reaction are a barium-141 nucleus and a krypton-92 nucleus. Three neutrons are also released: the original "bullet" neutron and 2 neutrons from the uranium nucleus.

The amount of energy released when a single uranium-235 nucleus splits is not very great. But the neutrons released in the first fission reaction become nuclear "bullets" that are capable of splitting other

---

represent the situation on the chalkboard; the letter was originally received (1), then sent to ten people (10), who each in turn sent the letter to ten different people (10 × 10 = 100). The letter was sent 1 + 10 + 100, or 111 times.)

• **Could this situation be called a chain reaction?** (Yes.)

• **What other kinds of chain reactions have you seen or heard about?** (Accept logical answers. Students might suggest a series of rear-end automobile collisions on an expressway during rush hour.)

CONTENT DEVELOPMENT

Point out that the nucleus of an atom stores a great amount of energy.

• **What is the name of the energy that holds a nucleus together?** (Have students recall that binding energy holds a nucleus together.)

• **Why must a nucleus be held together forcefully?** (Protons are positively charged, and like charges repel. The binding energy of the nuclear strong force "holds" the protons and neutrons together.)

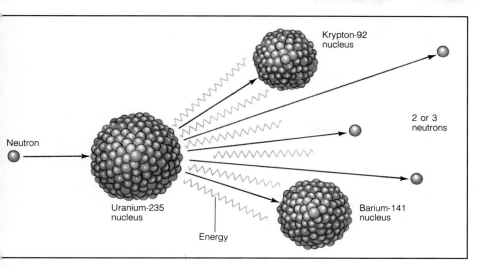

Figure 5–15 *In this diagram, a uranium-235 nucleus is bombarded with a neutron. The nucleus breaks up, producing a nucleus of krypton-92 and a nucleus of barium-141. Large amounts of energy as well as two additional neutrons are released. Each neutron is capable of splitting another uranium-235 nucleus. What is this repeating process called?* ②

uranium-235 nuclei. Each uranium nucleus that is split releases 3 neutrons. These neutrons may then split even more uranium nuclei. The continuous series of fission reactions is called a nuclear chain reaction. In a **nuclear chain reaction,** billions of fission reactions may take place each second!

When many atomic nuclei are split in a chain reaction, huge quantities of energy are released. This energy is produced as a result of the conversion of a small amount of mass into a huge amount of energy. The total mass of the barium, krypton, and 3 neutrons is slightly less than the total mass of the original uranium plus the initial neutron. The missing mass has been converted into energy. An uncontrolled chain reaction produces a nuclear explosion. The atomic bomb is an example of an uncontrolled chain reaction.

All currently operating nuclear power plants use controlled fission reactions to produce energy. The energy is primarily in the form of heat. The heat is carried away and used to produce electricity.

ACTIVITY READING

An Event That Would Change History Forever

With the excitement of science and the versatility of technology come serious responsibilities and social concerns. The development and use of the atomic bomb was one such ramification of the discovery of nuclear fission. Read *Hiroshima* by John Hersey to gain some insight into the effects such an event has had on people, politics, and history. ②

ACTIVITY READING

AN EVENT THAT WOULD CHANGE HISTORY FOREVER

Skill: Reading comprehension

You may want to have students prepare a written report or give a brief synopsis of the book to the class.

Integration: Use this Activity to integrate language arts into your lesson.

Explain that during nuclear fission, a neutron bombards a nucleus with such force that the binding energy is overcome and the nucleus splits into 2 smaller nuclei of approximately equal mass. This splitting of the nucleus releases large amounts of energy, as well as 2 additional neutrons.

• **What do you think would happen if these 2 neutrons bombarded 2 more nuclei?** (Accept reasonable answers. Lead students to infer that if other nuclei were in the vicinity, the 2 neutrons would bombard 2 additional nuclei, continuing the process, or reaction.)

Stress that this process or splitting is called a chain reaction. Mention that a nuclear fission reaction will continue until there are not sufficient numbers of nuclei available to continue the reaction.

• **Do you think nuclear fission occurs naturally, or must it be initiated and controlled artificially?** (The reaction must be initiated and controlled artificially.)

• **Why?** (Lead students to predict what such a naturally occurring fission scenario might be like.)

● ● ● ● **Integration** ● ● ● ●

Use the description of splitting uranium atoms to integrate concepts of nuclear fission into your lesson.

FISSION AND FUSION ENERGY

A 1.25-cm fuel pellet of uranium-235 can produce through fission as much energy as 800 kg of coal, 615 L of fuel oil, or 650 L of gasoline.

One liter of seawater, a common source of deuterium, contains the fusion energy potential of 300 L of gasoline.

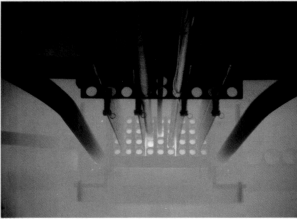

Figure 5–16 *Inside a nuclear reactor, radioactive fuel rods give off energy that produces a blue glow in the water. Before being placed in the reactor, a bundle of uranium-containing rods must be carefully checked.*

Nuclear Fusion

Another type of nuclear reaction that certain radioactive elements can undergo is called **nuclear fusion** (FYOO-zhuhn). Like fission, this kind of nuclear reaction produces a great amount of energy. But unlike fission, which involves the splitting of a high-mass nucleus, this reaction involves the joining of two low-mass nuclei. The word fusion means joining together. **Nuclear fusion is the joining of two atomic nuclei of smaller masses to form a single nucleus of larger mass.**

Nuclear fusion is a thermonuclear reaction. The prefix *thermo-* means heat. For nuclear fusion to take place, temperatures well over a million degrees Celsius must be reached. At such temperatures, the phase of matter known as plasma is formed. Plasma consists of positively charged ions, which are the nuclei of original atoms, and free electrons.

The temperature conditions required for nuclear fusion exist on the sun and on other stars. In fact, is nuclear fusion that produces the sun's energy. In the sun's core, temperatures of about 20 million degrees Celsius keep fusion going continuously. In a series of steps, hydrogen nuclei are fused into a helium-4 nucleus. See Figure 5–18.

Nuclear fusion produces a tremendous amount of energy. The energy comes from matter that is converted into energy during the reaction. In fact,

---

5–3 (continued)

ENRICHMENT

The amount of energy produced by the conversion of mass into energy is given by Einstein's equation

$$E = mc^2$$

where E is energy in ergs, m is mass in grams, and c is the speed of light, 3×10^{10} cm/sec. The squaring of the already large c term accounts mathematically for the enormous amounts of energy produced by conversions of even small quantities of mass.

CONTENT DEVELOPMENT

Remind students that fusion is the basic process that makes our sun shine. The conditions required for nuclear fusion exist on our sun and on other stars. Explain that the two conditions needed for nuclear fusion to occur are extreme heat and pressure. For fusion to occur, temperatures well over 1 million degrees must be present. Point out that these conditions do not exist naturally on Earth and that this is an important reason why scientists currently have considerable difficulties producing useful fusion reactions.
• **Do you think there is a limit to how long a fusion reaction can continue?** (Lead students to suggest yes.)

• **What do you think would limit the length of time of a fusion reaction?** (Accept logical answers. Students might suggest that the amount of fuel will limit the reaction—when the available fuel is spent the fusion reaction will stop.)

● ● ● ● **Integration** ● ● ● ●

Use the explanation of the prefix *thermo-* to integrate language arts into your science lesson.

the products formed by fusion have a mass that is about 1 percent less than the mass of the reactants. Although 1 percent loss of mass may seem a small amount, its conversion produces an enormous quantity of energy.

Nuclear fusion has several advantages over nuclear fission. The energy released in fusion reactions is greater for a given mass than that in fission reactions. Fusion reactions also produce less radioactive waste. And the possible fuels used for fusion reactions are more plentiful. Unfortunately, considerable difficulties exist with producing useful

Figure 5–17 *The light and heat that make life on Earth possible are the result of nuclear fusion within the sun. The destruction caused by a hydrogen bomb is also the result of nuclear fusion. The only source capable of delivering the energy required to trigger a fusion reaction in the hydrogen bomb is an atomic explosion.*

Figure 5–18 *In the process of nuclear fusion, hydrogen nuclei fuse to produce helium and tremendous amounts of energy. What other products are formed during the fusion reaction?* ❶

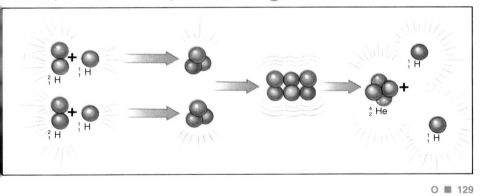

vantages over fission reactions: The energy released in fusion reactions is greater for a given mass than in fission reactions; fusion reactions produce less radioactive waste (although such waste is still produced), and the possible fuels for fusion reactions are more plentiful.

● ● ● ● **Integration** ● ● ● ●

Use the fact that nuclear fusion produces less radioactive waste than nuclear fission does to integrate ecology concepts into your lesson.

INDEPENDENT PRACTICE

▶ *Activity Book*

Students who need practice on the concept of harnessing the nuclei of atoms should complete the chapter activity Nuclear Fission and Nuclear Fusion. In this activity students will complete diagrams that illustrate the components of nuclear fission and nuclear fusion.

ENRICHMENT

In the laboratory, scientists try to reproduce "star power," or the fusion reaction of our sun, by combining two forms of hydrogen—deuterium and tritium. Have interested students consult up-to-date library sources to find out about current attempts to produce energy by nuclear fusion. Students might also choose to draw diagrams of the various structures used and proposed by scientists and then share their work with the class.

CONTENT DEVELOPMENT

You might choose to explain fusion by making use of nuclear equations. Contrast nuclear fusion and nuclear fission, stressing the high-energy conditions that are necessary for fusion to occur.

Refer students to the fusion reaction shown in Figure 5–18 and have them compare it with the fission reaction shown in Figure 5–15. As each diagram is examined, lead students to understand that fusion reactions contain several distinct ad-

INTEGRATION
FOOD SCIENCE

Radioisotopes are commonly used in Europe to sterilize food. Such treated foods, including milk, can be stored for months without refrigeration. These sterilized foods contain no radioactive materials.

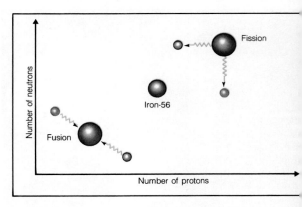

Figure 5–19 *Different elements will give off energy and become more stable by undergoing either fusion or fission. For most elements lighter than iron-56, nuclear fusion will give off energy. For most elements heavier than iron-56, nuclear fission will give off energy. Elements gain stability by moving closer on the graph to iron-56.*

fusion reactions on Earth. Fusion reactions are more difficult to begin, to control, and to maintain than nuclear fission reactions are. After all, no known vessel can contain reactions occurring at such tremendous temperatures. And such high temperatures are extremely difficult to achieve. In fact, a hydrogen bomb, which uses fusion, is started by an atomic bomb, which uses fission. It is the only way of achieving the necessary temperatures.

Scientists are continuing their search for ways to control this powerful reaction and to tap a tremendous energy resource. As an example, experiments using high-powered laser beams and electrons as ways of starting fusion reactions are being conducted

5–3 Section Review

1. What is nuclear fission? Nuclear fusion?
2. How is the sun's energy produced?
3. Where does the energy produced in both fission and fusion reactions come from?
4. Compare the energy produced by fission and fusion reactions with the energy produced by radioactive decay.

Connection—*You and Your World*
5. Why is the production of less radioactive waste considered an advantage of nuclear fusion?

130 ■ O

5–4 Detecting and Using Radioactivity

Radioactivity cannot be seen or felt. Becquerel discovered radioactivity because it left marks on photographic film. Although film is still used today to detect radioactivity, scientists have more specialized instruments for this purpose. **The instruments scientists use to detect and measure radioactivity include the electroscope, the Geiger counter, the cloud chamber, and the bubble chamber.**

Instruments for Detecting and Measuring Radioactivity

ELECTROSCOPE An **electroscope** is a simple device that consists of a metal rod with two thin metal leaves at one end. If an electroscope is given a negative charge, the metal leaves separate. In this condition, the electroscope can be used to detect radioactivity.

Radioactive substances remove electrons from molecules of air. As a result, the molecules of air become positively charged ions. When a radioactive substance is brought near a negatively charged electroscope, the air molecules that have become positively charged attract the negative charge on the leaves of the electroscope. The leaves discharge, or lose their charge, and collapse.

GEIGER COUNTER In 1928, Hans Geiger designed an instrument that detects and measures radioactivity. Named the **Geiger counter** in honor of its inventor, this instrument produces an electric current in the presence of a radioactive substance.

A Geiger counter consists of a tube filled with a gas such as argon or helium at a reduced pressure. When radiation enters the tube through a thin window at one end, it removes electrons from the atoms of the gas. The gas atoms become positively charged ions. The electrons move through the positively charged ions to a wire in the tube, setting up an electric current. The current, which is amplified and fed into a recording or counting device, produces a flashing light and a clicking sound. The number of

Figure 5–20 *Because radioactive substances will cause an electroscope to discharge, an electroscope can be used to detect radiation. What do radioactive substances do to molecules of air?* ❶

O ■ 131

5–4 Detecting and Using Radioactivity

MULTICULTURAL OPPORTUNITY 5–4

Have students investigate the life and work of Dr. Dixie Lee Ray. Dr. Ray began her career as a marine biologist, but by combining her interest in theater she became a leader in promoting public understanding of science. In 1973, she was appointed the chairperson of the Atomic Energy Commission. Three years later, she was elected the first woman governor of the state of Washington.

ESL STRATEGY 5–4

Point out that the word *leaves* can have different meanings. It can be used to describe parts of an electroscope, and it also can be used in a description of a cloud chamber and a bubble chamber. The word also refers to the leaves of a plant. Have students write three sentences demonstrating that they understand the different meanings of *leaves*.

TEACHING STRATEGY 5–4

FOCUS/MOTIVATION

Use a simple laboratory electroscope to illustrate some of the electrical principles involved in this section of the textbook. Rub a rubber rod with fur to give the electroscope a negative charge. Bringing a positively charged glass rod (which has been rubbed with silk) near the knob of the electroscope will then bring about discharge, which is what would occur if a radioactive substance were brought near.

CONTENT DEVELOPMENT

Discuss the structure, operation, and use of electroscopes such as the one used in the Focus/Motivation activity.

● ● ● ● **Integration** ● ● ● ●

Use the explanation of how radioactive substances remove electrons from air molecules to integrate charge concepts into your lesson.

---

things. Regardless of where radioactive waste is located, it will emit radioactive energy for many thousands or millions of years. Because radioactive waste is such a long-term hazard, it is better to produce less of it than more of it, given the choice.

REINFORCEMENT/RETEACHING

Monitor students' responses to the Section Review questions. If students appear to have difficulty with any of the questions, review the appropriate material in the section.

CLOSURE

▶ *Review and Reinforcement Guide*

At this point have students complete Section 5–3 in the *Review and Reinforcement Guide.*

More than 100 radioactive isotopes have been used in the field of medicine.

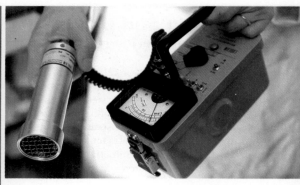

Figure 5–21 *A Geiger counter detects and measures radioactivity. A spiderwort plant is nature's radiation detector. The stamens of the spiderwort flower are usually blue. In the presence of radiation the stamens turn pink.*

Figure 5–22 *This cloud chamber photograph shows an upward stream of alpha particles.*

flashes and clicks per unit time indicates the strength of the radiation. A counter attached to the wire is able to measure the amount of radioactivity by measuring the amount of current.

CLOUD CHAMBER   A **cloud chamber** contains a gas cooled to a temperature below its usual condensation point (point at which it becomes a liquid). When a radioactive substance is put inside the chamber, droplets of the gas condense around the radioactive particles. The process is similar to what happens in "cloud seeding," when rain droplets condense around particles that have been injected into the clouds. The droplets formed around the particles of radiation in a cloud chamber leave a trail that shows up along the chamber lining. An alpha particle leaves a short, fat trail, whereas a beta particle's trail is long and thin.

BUBBLE CHAMBER   The **bubble chamber** is similar in some ways to the cloud chamber, although its construction is more complex. A bubble chamber contains a superheated liquid. A superheated liquid is hot enough to boil—but does not. Instead, it remains in the liquid phase. The superheated liquid most often contained in a bubble chamber is hydrogen.

When radioactive particles pass through the chamber, they cause the hydrogen to boil. The boiling liquid leaves a trail of bubbles, which is used to track the radioactive particle. The photograph you saw at the beginning of this chapter is of a bubble chamber after a nuclear reaction.

5–4 (continued)

CONTENT DEVELOPMENT

Obtain a Geiger counter and use it to illustrate that nuclear decay is occurring, in small amounts, at all times. Nuclear decay is also measurable in terms of background radiation. You might also wish to obtain a small amount of a relatively safe radioactive substance, such as $^{131}_{53}I$, from a chemical supply house and demonstrate the ability of the Geiger counter to detect radiation produced by beta decay. Measurements of the numbers of counts conducted at intervals over a series of days can be represented on a graph to determine the approximate half-life of this isotope, which is about eight days. **Note:** *Follow all safety precautions suggested by the supplier.*

● ● ● ●   **Integration**   ● ● ● ●

Use Figure 5–21 and the spiderwort's ability to detect radiation to integrate botany concepts into your lesson.

GUIDED PRACTICE

Skills Development
Skill: Relating concepts

After students receive an explanation of the structure, operation, and use of bubble chambers and cloud chambers, help them reinforce their understanding of radiation-detection instruments by allowing them to work in small groups and develop questions about each instrument. Sample questions might include these: In which radiation detection device do droplets of gas condense around radioactive particles? How does a gas-filled tube in a Geiger counter produce electrical current? A list of questions should then be exchanged between groups.

CONTENT DEVELOPMENT

Remind students that they have already discussed two practical applications of radioactive substances. One application was the use of the radioactive isotope carbon-14 and uranium-238 to determine the approximate age of fossils or other objects.

Putting Radioactivity to Work

Radioactive substances have many practical uses. Dating organic objects, which you learned about earlier, is one such use. In industry, radioactive isotopes, or **radioisotopes,** have additional uses. Radioisotopes can be used to find leaks or weak spots in metal pipes, such as oil pipe lines. Radioisotopes also help study the rate of wear on surfaces that rub together. One surface is made radioactive. Then the amount of radiation on the other surface indicates the wear.

Because radioisotopes can be detected so readily, they can be used to follow an element through an organism, or through an industrial process, or through the steps of a chemical reaction. Such a radioactive element is called a **tracer,** or radiotracer. Tracers are possible because all isotopes of the same element have essentially the same chemical properties. When a small quantity of radioisotope is mixed with the naturally occurring stable isotopes of the same element, all the isotopes go through the same reactions together.

An example of a tracer is phosphorus-32. The nonradioactive element phosphorus is used in small amounts by both plants and animals. If phosphorus-32 is given to an organism, the organism will use the radioactive phosphorus just as it does the nonradioactive phosphorus. However, the path of the radioactive element can be traced. In this way, scientists can learn a great deal about how plants and animals use phosphorus.

Another area in which radioisotopes make an important contribution is in the field of medicine. The branch of medicine in which radioactivity is used is known as nuclear medicine. Tracers are extremely valuable in diagnosing diseases. For example, radioactive iodine—iodine-131—can be used to study the function of the thyroid gland, which absorbs iodine. Sodium-24 can be used to detect diseases of the circulatory system. Iron-59 can be used to study blood circulation.

Another procedure, known as radioimmunoassay, developed by Dr. Rosalyn Yalow—who won the Nobel Prize for her work—involves using tracers to detect the presence of minute quantities of substances in

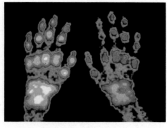

Figure 5–23 *Radioisotopes can be used as tracer elements to produce images such as this one of a person's hands. The different-colored areas help doctors diagnose conditions in order to prescribe treatment. This person has arthritis.*

ACTIVITY
DOING

Radioactivity in Medicine

Obtain permission to visit the radiation laboratory at a local hospital. Observe the instruments used and the safety precautions taken for both patients and technicians when using the devices. Interview a technician, if possible. Make a report on your visit.

③

ACTIVITY
DOING
RADIOACTIVITY IN MEDICINE

Skills: Making observations, relating concepts

This field investigation will allow students to relate some of the topics and concepts covered in the textbook to various lifesaving applications. If students are able to visit a hospital, encourage them to take notes and to report their findings to the class. If a visit is not possible, invite a medical technician from a local hospital (contact the public relations department) to class to discuss radioactivity and its applications in medicine.

Integration: Use this Activity to integrate language arts skills into your science lesson.

Another application was the discovery and creation of transuranium elements—elements that did not previously exist.

Explain to students that radioactive substances have many other practical applications. Tracers are extremely valuable in the field of nuclear medicine. Tracers have made it possible to diagnose various diseases that previously were much more difficult to diagnose. Also, experiments were undertaken in the late 1950s that used tracers to determine precisely which compounds were produced by plants during the process of photosynthesis.

Another application of radioactive substances involves studying the change from stable isotopes to radioactive isotopes of a sample bombarded by neutrons. This process, known as neutron activation analysis, has been used to match hair samples—one found at a crime scene and one from a suspect's head, for example—and to determine the authenticity of oil paintings. Because the pigments used in oil paintings hundreds of years old are different from the pigments used in oil paintings today, neutron activation analysis can be used to determine authentic paintings from those that have been "faked."

● ● ● ● **Integration** ● ● ● ●

Use the role of tracers in the diagnosis of disease to integrate concepts of medicine into your science lesson.

RADIOISOTOPES

The use of radioisotopes has significantly influenced the field of medicine and the quality of health care its workers and researchers can provide. The first medical use of radioisotopes occurred in the treatment of cancer.

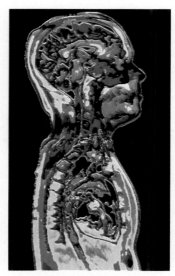

Figure 5–24 *Observing changes in the nuclei of atoms has many important applications. Through nuclear magnetic resonance imaging, doctors can form detailed images of various parts of the human body. What is an advantage of this procedure over exploratory surgery?* ❶

the body. These tests can be used to detect pregnancy as well as the early signs of a disease. Another powerful research tool—nuclear magnetic resonance imaging (MRI)—has become invaluable in a variety of fields, from physics to chemistry and biochemistry. MRI involves recording changes in the energy of atomic nuclei in response to external energy changes, without altering the cells of the body in any way.

Radiation is also used to destroy unhealthy cells, such as those that cause cancer. Radiation in large doses destroys living tissues, especially cells undergoing division. Because cancer cells undergo division more frequently than normal cells do, radiation kills more cancer cells than it does normal cells. As early as 1904, physicians attempted to treat masses of unhealthy cells, known as tumors, with high-energy radiation. This treatment is called radiation therapy.

Radioisotopes can also be used to kill bacteria that cause food to spoil. Radiation was used to preserve the food that the astronauts ate while on the moon and in orbit.

Dangers of Radiation

Although radioactivity has tremendous positive potential, radioactive materials must be handled with great care. Radioactive materials are extremely dangerous. Radiation can ionize—or knock electrons out of—atoms or molecules of any material it passes through. For this reason, the term ionizing radiation is sometimes used. So, oddly enough, the same radiation that is used to treat disease can also cause it.

Ionization can cause considerable damage to materials, particularly to biological tissue. When ionization is produced in cells, ions may take part in chemical reactions that would not otherwise have occurred. This may interfere with the normal operation of the cell. Damage to DNA is particularly serious. An alteration in the DNA (substance responsible for carrying traits from one generation to another) of a cell can interfere with the production of proteins and other essential cellular materials. The result may be the death of the cell. If many cells die, the organism may not be able to survive.

CONTENT DEVELOPMENT

Discuss other practical applications of radioactivity such as magnetic resonance imaging, radiation therapy, and food processing. Have students describe other uses of radioactivity that are not specifically mentioned in the textbook. Also, allow students an opportunity to discuss the hazards of radioactive substances and to speculate about other potential applications of radioactivity in the future.

● ● ● ● **Integration** ● ● ● ●

Use the applications of magnetic resonance imaging to integrate MRI concepts into your lesson.

Use the relationship between radioactivity and cell damage to integrate concepts of genetics into your lesson.

GUIDED PRACTICE

Skills Development

Skill: Interpreting illustrations

Have students collect from books and magazines photographs and advertisements that illustrate uses of radioisotopes. Display or circulate the collection and ask:

• **What radioisotope applications are being illustrated?** (Answers will vary, depending on specific illustrations.)
• **How are isotopes being used?** (Accept correct answers.)

Remind students of the potential dangers of using radioactivity and radioactive substances. Ask:

• **What are some potential dangers of the radioactive applications in these pictures?** (Accept reasonable answers.)

Encourage the entire class to debate the issue of whether the practical uses of radioactive substances outweigh the dangers involved.

REINFORCEMENT/RETEACHING

For each application of isotopes, point out the property of radioactive substances that makes the application possible. For example, because radioisotopes can decay over extended periods of time and the products of this decay are detectable, they are used as tracers. Because they release highly energetic particles and large amounts of energy that can destroy living

Large doses of radiation can cause reddening of the skin, a drop in the white blood cell count, and numerous other unpleasant symptoms, including nausea, fatigue, and loss of hair. Such effects are sometimes referred to as radiation sickness. Large doses of radiation can also be fatal. Marie Curie's death in 1934 was caused by exposure to too much radiation.

Even metals and other structural materials can be weakened by intense radiation. This is a considerable problem in nuclear-reactor power plants and for space vehicles that must pass through areas of intense cosmic radiation.

We are constantly exposed to low-level radiation from natural sources such as cosmic rays from space, radioactivity in rocks and soil, and radioactive isotopes that are present in food and in our bodies.

Today, people who work with radioactive materials take extreme precautions. They wear radiation-sensitive badges that serve as a warning of unsafe levels of radiation. Specially designed clothing is worn to block radiation. Scientists continue to search for greater understanding and control of radiation so that its benefits can be enjoyed without the threat of danger.

Figure 5–25 *The hands and numbers of a luminous watch contain minute amounts of radium. The radioactive decay of radium causes the watch to glow in the dark. People who painted the dials on clocks in the early 1900s suffered from radiation poisoning because they often licked the tips of their brushes to make fine lines.*

Figure 5-26 *Workers handling radioactive materials must use extreme caution and wear specially designed clothes for protection.*

5–4 Section Review

1. Name four instruments used to detect and measure radioactivity.
2. Compare a bubble chamber and a cloud chamber.
3. What is a radioisotope? What are its uses?
4. What is a tracer? Describe several uses of tracers.

Connection—*Life Science*

5. A mutation is a change that occurs in the genetic material of a cell (the code that determines what traits will be carried on from generation to generation). Why does nuclear radiation cause mutations?

tissue and organisms, they are used in cancer treatment and in food sterilization.

INDEPENDENT PRACTICE
Section Review 5–4

1. Four instruments used to detect radioactivity include the electroscope, Geiger counter, bubble chamber, and cloud chamber.

2. In both a bubble chamber and a cloud chamber, a radioactive substance leaves a trail. In a cloud chamber, droplets form around a radioactive substance and leave a trail along the chamber lining. In a bubble chamber, the trail of a radioactive substance consists of bubbles created from boiling liquid hydrogen.

3. A radioisotope is an artificially produced radioactive isotope of a common element. Radioisotopes are used to study living organisms, to diagnose and treat diseases, to sterilize foods, and to monitor industrial processes. A tracer is a radioactive element whose path can be followed through various steps of a chemical reaction. Tracers are used to study certain internal processes of living things, to diagnose diseases, and to detect minute quantities of substances in the human body.

5. Radiation destroys, changes, or rearranges some or all of the genetic instructions in the hereditary material of a cell.

REINFORCEMENT/RETEACHING

Review students' responses to the Section Review questions. Reteach any material that is still unclear, based on students' responses.

CLOSURE

▶ *Review and Reinforcement Guide*
Students may now complete Section 5–4 in the *Review and Reinforcement Guide*.

Laboratory Investigation

THE HALF-LIFE OF A SUGAR CUBE

BEFORE THE LAB

1. Gather all materials at least one day prior to the investigation. You should gather enough materials to meet your class needs, assuming two to six students per group.

2. Test-mark a sugar cube with food coloring to make certain that the coloring does not soak through the cube and make it difficult for the marked face to be distinguished from the unmarked faces.

PRE-LAB DISCUSSION

Have students read the complete laboratory procedure.

Before beginning the actual investigation, allow students to examine a sugar cube.

• **How many different faces does a sugar cube have?** (Six.)

• **What is the probability that a sugar cube will land with a given face up, using a random toss?** (One in six, assuming a regular cube with equal density throughout.)

Ask students to formulate a hypothesis regarding the number of tosses required to remove one half of the cubes. Review the concept of half-life and make sure students recognize the analogy between a radioactive-decay half-life situation and the sugar cube situation they are investigating.

Laboratory Investigation

The Half-Life of a Sugar Cube

Problem

How can the half-life of a large sample of sugar cubes be determined?

Materials (per group)

250 sugar cubes	large bowl
food coloring	medicine dropper

Procedure

1. Place a small drop of food coloring on one side of each sugar cube.

2. Put all the sugar cubes in a bowl. Then gently spill them out on the table. Move any cubes that are on top of other cubes.

3. Remove all the sugar cubes that have the colored side facing up. If you have room on the table, arrange in a vertical column the sugar cubes that you removed. Put the rest of the cubes back in the bowl.

4. Repeat step 3 several more times until five or fewer sugar cubes remain.

5. On a chart similar to the one shown, record the number of tosses (times you spilled the sugar cubes), the number of sugar cubes removed each time, and the number of sugar cubes remaining. For example, suppose after the first toss you removed 40 sugar cubes. The number of tosses would be 1, the number of cubes removed would be 40, and the number of cubes remaining would be 210 (250 – 40).

Observations

1. Make a full-page graph of tosses versus cubes remaining. Place the number of tosses on the X (horizontal) axis and the number of cubes remaining on the Y (vertical) axis. Start at zero tosses with all 250 cubes remaining.

136 ■ O

2. Determine the half-life of the decaying sugar cubes in the following way. Find the point on the graph that corresponds to one half of the original sugar cubes (125). Move vertically down from this point until you reach the horizontal axis. Your answer will be the number of tosses.

Tosses	Sugar Cubes Removed	Sugar Cubes Remaining
0	0	250
1	40	210
2		
3		

Analysis and Conclusions

1. What is the shape of your graph?

2. How many tosses are required to remove one half of the sugar cubes?

3. How many tosses are required to remove one fourth of the sugar cubes?

4. Assuming tosses are equal to years, what is the half-life of the sugar cubes?

5. Using your answer to question 4, how many sugar cubes should remain after 8 years? After 12 years? Do these numbers agree with your observations?

6. What factor(s) could account for the differences in your observed results and those calculated?

7. **On Your Own** Repeat the experiment with a larger number of sugar cubes. Predict whether the determined half-life will be different. Is it?

TEACHING STRATEGIES

1. Encourage students not to "look" while making their tosses—reminding them to toss the cube in such a way that the number of sides with coloring appears randomly as often as possible throughout each toss.

2. Remind students that the best results will be achieved if the sugar cubes are not broken or chipped in any way. Remind students to toss the cubes gently.

DISCOVERY STRATEGIES

Discuss how the investigation relates to the chapter by asking open questions similar to the following:

• **Why are sugar cubes used in this investigation in place of genuine radioactive materials which have an actual half-life that could be determined?** (Radioactive materials emit energy that is known to cause genetic cell damage and sometimes death—inferring, relating.)

Study Guide

Summarizing Key Concepts

5–1 Radioactivity

▲ An element that gives off nuclear radiation is said to be radioactive.

▲ Nuclear radiation occurs in three forms: alpha particles, beta particles, and gamma rays.

5–2 Nuclear Reactions

▲ If the binding energy—the force that holds the nucleus together—is not strong, an atom is said to be unstable. Atoms with unstable nuclei are radioactive.

▲ Atoms that have the same atomic number but different numbers of neutrons are called isotopes.

▲ An unstable nucleus eventually becomes stable by undergoing a nuclear reaction.

▲ Radioactive decay is a nuclear reaction that involves the spontaneous breakdown of an unstable nucleus. During radioactive decay, alpha particles, beta particles, and/or gamma rays are emitted.

▲ The decay of a radioactive element occurs at a fixed rate called the half-life.

▲ The series of steps by which a radioactive nucleus decays is known as a decay series.

▲ Artificial transmutation involves the nuclear reactions in which atomic nuclei are bombarded with high-speed particles.

5–3 Harnessing the Nucleus

▲ Nuclear fission is the splitting of an atomic nucleus to form two smaller nuclei of roughly equal mass.

▲ Nuclear fusion is the joining together of two atomic nuclei to form a single nucleus of larger mass.

5–4 Detecting and Using Radioactivity

▲ Four devices that can detect radioactivity are the electroscope, Geiger counter, cloud chamber, and bubble chamber. The Geiger counter can also measure radioactivity.

▲ Radioactive substances must be handled carefully because large amounts of radiation can be harmful to living things.

Reviewing Key Terms

Define each term in a complete sentence.

5–1 Radioactivity
nuclear radiation
radioactive
radioactivity
alpha particle
beta particle
gamma ray

5–2 Nuclear Reactions
nuclear strong force
binding energy

isotope
radioactive decay
transmutation
half-life
decay series
artificial transmutation
transuranium element

5–3 Harnessing the Nucleus
nuclear fission

nuclear chain reaction
nuclear fusion

5–4 Detecting and Using Radioactivity
electroscope
Geiger counter
cloud chamber
bubble chamber
radioisotope
tracer

ANALYSIS AND CONCLUSIONS

1. Students may describe their graph using words or by drawing its shape. The ideal shape would display a hyperbolic curve in the first quadrant, asymptotic with respect to the X axis.

2. Answers will vary, but approximately four tosses should remove one half of the cubes.

3. Approximately $1\frac{1}{2}$ to 2 tosses should remove one fourth of the cubes.

4. Four years.

5. One fourth; one eighth. Students' data should be similar, but probably not exact as the ideal figures given.

6. Accept all logical answers. Half-life is a statistical measurement and cannot be considered exact using small sample sizes. As such, results from group to group, as well as from group to ideal, will vary.

7. The half-life should not change considerably when more cubes are added. This can be inferred because the half-life of an element is the same regardless of how much of that element is considered.

GOING FURTHER: ENRICHMENT
Part 1

Have interested students repeat the investigation using a larger number of sugar cubes.

Part 2

Ask students to predict the result of three investigations identical to the investigation just performed except for the use of 4-faced (tetrahedral), 12-faced (dodecahedral), and 20-faced (icosahedral) regular solids instead of the 6-faced (cubic) regular solids.

• **When you are exposed to radioactive materials such as those used in a dentist's X-ray machine, is it important to protect areas of your body other than your mouth from receiving radiation?** (Yes—relating, concluding.)

• **Why?** (Radioactivity causes damage to cells and genetic material—relating, applying.)

• **How can you protect other areas of your body while receiving a mouth X-ray?** (Ask to wear a lap shield or a lead apron, which will help to protect other areas of the body from radioactivity exposure—applying, relating.)

OBSERVATIONS

1. Check students' graphs to see that each graph accurately reflects the data.

2. Check students' determination of half-life to see that it reflects their graphed data.

Chapter Review

ALTERNATIVE ASSESSMENT

The *Prentice Hall Science* program includes a variety of testing components and methodologies. Aside from the Chapter Review questions, you may opt to use the Chapter Test or the Computer Test Bank Test in your *Test Book* for assessment of important facts and concepts. In addition, Performance-Based Tests are included in your *Test Book*. These Performance-Based Tests are designed to test science process skills, rather than factual content recall. Since they are not content dependent, Performance-Based Tests can be distributed after students complete a chapter or after they complete the entire textbook.

CONTENT REVIEW

Multiple Choice

1. d
2. a
3. a
4. b
5. c
6. b
7. d
8. c

True or False

1. F, polonium/radium
2. T
3. T
4. F, radioactive decay
5. F, fusion
6. F, Geiger counter
7. T
8. T

Concept Mapping

Row 1: Nuclear radiation
Row 2: Radioactive element; Beta
   particles; Gamma rays

CONCEPT MASTERY

1. Becquerel observed something unusual coming from uranium salt, and he wanted to find out what was producing this "something" that is now called radiation. He hypothesized that it was coming from the uranium and subsequently tested many uranium compounds. Each compound produced radiation. Based on his observations, Becquerel concluded that uranium was responsible for the radiation.
2. An alpha particle can be stopped by a sheet of paper. A beta particle can pass through as much as 3 millimeters of aluminum. A gamma ray can penetrate several centimeters of lead.
3. An understanding of what makes a nucleus break apart and the reasons why only some elements decay led to an understanding of the strong force. The energy associated with the strong force is binding energy, and this energy overcomes the desire of protons to repel because of their like charges.
4. Isotopes of a given element contain the same number of protons but different numbers of neutrons.
5. Radioactive decay: the spontaneous breakdown of an unstable atomic nucleus, causing the nucleus to become lighter and emit particles or rays. Artificial transmutation: the bombarding of nuclei with high-speed particles, creating elements that have atomic numbers greater than 92. Nuclear fission: the splitting of an atomic nucleus into 2 smaller nuclei of approximately equal mass, which initiate a chain reaction that releases tremendous

---

Chapter Review

Content Review

Multiple Choice

Choose the letter of the answer that best completes each statement.

1. The particle given off by a radioactive element that is actually a helium nucleus is a(an)
   a. beta particle.   c. isotope.
   b. gamma ray.   d. alpha particle.
2. Atoms with the same atomic number but different numbers of neutrons are
   a. isotopes.   c. alpha particles.
   b. radioactive.   d. beta particles.
3. In relation to the original atom, the atom that results from alpha decay has an atomic number that is
   a. 2 less.   c. the same.
   b. 1 less.   d. 2 more.
4. The process in which one element changes into another as a result of nuclear changes is
   a. fluorescence.   c. transuranium.
   b. transmutation.   d. synthesis.
5. An atomic nucleus splits into two smaller nuclei in
   a. fusion.   c. fission.
   b. alpha decay.   d. transmutation.
6. A device in which radioactive materials leave a trail of liquid droplets is a(an)
   a. bubble chamber.   c. decay chamber.
   b. cloud chamber.   d. electroscope.
7. A Geiger counter detects radioactivity when the radioactive substance
   a. leaves a trail of bubbles.
   b. condenses around particles of a gas.
   c. causes liquid gas to boil.
   d. produces an electric current.
8. An artificially produced radioactive isotope of an element is called a
   a. synthetic isotope.   c. radioisotope.
   b. transmutation.   d. gamma isotope.

True or False

If the statement is true, write "true." If it is false, change the underlined word or words to make the statement true.

1. One of the radioactive elements discovered by the Curies was <u>uranium</u>.
2. The energy that holds the nucleus together is the <u>binding energy</u>.
3. <u>Gamma rays</u> are electromagnetic waves of very high frequency and energy.
4. The spontaneous breakdown of an unstable nucleus is <u>artificial transmutation</u>.
5. Two atomic nuclei join together during nuclear <u>fission</u>.
6. Gas molecules are ionized by a radioactive substance in a <u>bubble chamber</u>.
7. <u>Iodine-131</u> can be used to study the function of the thyroid gland.
8. <u>Radioisotopes</u> are used to kill bacteria.

Concept Mapping

Complete the following concept map for Section 5–1. Refer to pages O6–O7 to help you construct a concept map for the entire chapter.

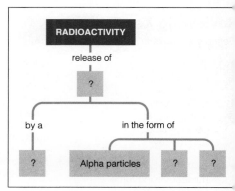

Concept Mastery

Discuss each of the following in a brief paragraph.

1. Describe how Becquerel's work illustrates the scientific method.
2. Describe the penetrating power of each of the three types of radiation.
3. How was the existence of the nuclear strong force deduced? What is the energy associated with the strong force, and how is it related to the stability of an element?
4. What do isotopes of a given element have in common? How are they different?
5. Describe the four types of nuclear reactions.
6. Describe how a Geiger counter works. How is it different from a bubble chamber, cloud chamber, and an electroscope?
7. Explain why the amount of helium in the sun is increasing.
8. Why must the fission process release neutrons if it is to be useful?
9. What are some of the uses of radioactivity?

Critical Thinking and Problem Solving

Use the skills you have developed in this chapter to answer each of the following.

1. **Making graphs** Sodium-24 has a half-life of 15 hours. Make a graph to show what happens to a 100-gram sample of sodium-24 over a 5-day period.
2. **Interpreting a graph** The dots on the accompanying graph represent stable nuclei. The straight line represents equal numbers of protons and neutrons. Describe the relationship between the number of protons and the number of neutrons as the atomic number increases.

3. **Making comparisons** Describe the three types of radiation and relate them to the three types of radioactive decay.
4. **Analyzing data** A skeleton of an ancient fish is found to contain one eighth the amount of carbon-14 that it contained when it was alive. How old is the skeleton?
5. **Making calculations** The half-life of cobalt-60 is 5.26 years. How many grams of a 20-gram sample of cobalt-60 remain after 10.52 years? After 15.78 years?
6. **Drawing conclusions** Why do you think it is more dangerous for a woman of child-bearing age to be exposed to nuclear radiation than it is for a younger or an older woman to be exposed?
7. **Using the writing process** Food spoils because organisms such as bacteria and mold grow on it and decompose it. Gamma rays can destroy such organisms. So foods treated with gamma rays, or irradiated foods, last longer. Many people, however, are afraid that irradiated foods may be dangerous. Others do not want to live near gamma-ray treatment centers. Write a letter to your state representative describing your thoughts about irradiated foods.

---

8. The energy of fission requires a sustained chain reaction. Released neutrons function as bullets to continue the reaction.
9. Uses will vary. Students might suggest radioactivity is used in medicine to fight and to diagnose certain diseases, to sterilize foods, and to produce power.

CRITICAL THINKING AND PROBLEM SOLVING

1. Check students' graphs to make sure that the plotting is consistent with the data provided.
2. The number of neutrons increases at a rate faster than the rate at which the number of protons increases.
3. Alpha decay: The nucleus of a radioactive atom releases 2 protons and 2 neutrons, or an alpha particle; the atomic number of the new element formed is two less than the original atomic number. Beta decay: An electron is released from the nucleus during beta decay; the element produced has the same mass number as the original atom with an atomic number one greater than before. Gamma decay: Gamma rays are released; gamma decay usually accompanies alpha and beta decay.
4. Assuming that the half-life of carbon is 5730 years, the skeleton is 17,190 years old.
5. 5 grams; 2.5 grams.
6. The woman's hereditary material inside her egg cells may be damaged or destroyed by her exposure to nuclear radiation.
7. Letters will vary, but they should state a position with respect to irradiated foods.

KEEPING A PORTFOLIO

You might want to assign some of the Concept Mastery and Critical Thinking and Problem Solving questions as homework and have students include their responses to unassigned questions in their portfolio. Students should be encouraged to include both the question and the answer in their portfolio.

ISSUES IN SCIENCE

The following issue can be used as a springboard for discussion or given as a writing assignment.

1. Fission reactions can be carried out in nuclear power plants to produce energy. However, there are dangers associated with such reactors. Do you think the advantages outweigh the disadvantages?

---

amounts of heat and energy. Nuclear fusion: the joining of 2 atomic nuclei of smaller masses to form a single nucleus of larger mass, occurring in conditions of great temperature and pressure and releasing tremendous amounts of energy.
6. In a Geiger counter, radiation removes electrons from the atoms of a gas, and the resulting positive ions create a current that is read. An electroscope is given a negative charge and will react to a radioactive substance because radioactive substances remove electrons from the air,

creating positively charged ions. In a cloud chamber, droplets of gas condense around radioactive particles, leaving a trail along the chamber lining. In a bubble chamber, a radioactive particle causes hydrogen to boil, leaving a trail of bubbles.
7. The sun's energy is produced during the fusion process. In this process hydrogen nuclei fuse to form helium nuclei. Each instant of its life, the sun consumes hydrogen and produces helium.

ORGANIC BALLS: NEW ADVENTURES IN CHEMISTRY

Background Information

Because of its soccer-ball shape, which resembles the geodesic dome structures created by Buckminster Fuller, Smalley and Kroto whimsically named the carbon molecules they discovered Buckminsterfullerenes. Smalley and Kroto believe that bucky ball molecules are everywhere on Earth—in the air, on land, and in the sea—and that they might be one of the most abundant molecules in the universe. Buckminsterfullerene, or carbon-60, is very unreactive. Once a molecule of carbon grows to the size of the bucky ball, it tends to stay there. Because it is spherical without loose ends, it does not attract other atoms or molecules.

Smalley and Kroto hypothesize that fullerenes are formed when soot is formed. They believe that fullerenes begin as a few carbon atoms joined in sheets, with dangling molecular bonds. Other carbon atoms are attracted and attach themselves, causing the sheets to curve and finally close off. Because temperature, gas composition, fuel chemistry, and other factors affect the formation of fullerenes, it would be rare for naturally occurring fullerenes to form. Other scientists disagree with Smalley and Kroto. Research continues, and scientists are working to support the theory of fullerene's natural existence and to develop uses for this remarkably stable molecule.

ORGANIC BALLS:
New Adventures In Chemistry

You can't kick it through goalposts, dribble it down a basketball court, or hit it with a bat or a tennis racket. But scientists are extremely excited about a ball that has much more to do with atoms than with athletics!

The special sphere is called the "bucky ball." It is a molecular form of pure carbon shaped like the geodesic domes designed by the famous architect R. Buckminster Fuller. Thus its nickname—bucky ball—and its complete name: Buckminsterfullerene.

The humorous scientists who named the fullerene—as well as discovered it—include teams of chemists led by Richard Smalley of Rice University in Texas and Harry Kroto of the University of Sussex in England. In 1985, while researching the clustering properties of carbon atoms, the chemists noticed what they considered to be a curious trend: All the carbon molecules they produced contained an even number of atoms. Their measurements also indicated that a large proportion of the molecules consisted of 60 atoms. After further investigation—with equipment that ranged from sophisticated lasers to simple scraps of paper—the scientists theorized the existence of a special crystal form of carbon. Under the right circumstances, they claimed, 60 carbon atoms could bond together to form a hollow perfect sphere: the bucky ball. The sphere has 32 facets, or sides, of which 12 are pentagons (5-sided) and 20 are hexagons (6-sided). Its structure makes it an incredibly stable molecule.

Smalley and Kroto were offering the world a startling discovery: a third molecular form of pure carbon. The only other known forms of carbon are graphite and diamond. In light of its implications, Smalley and Kroto's proposal was first met with skepticism from

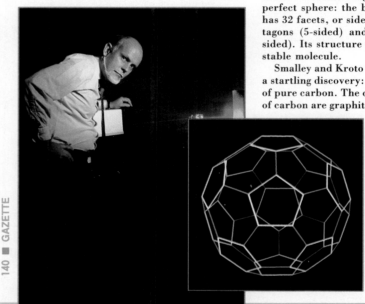

◀ **Dr. Richard Smalley is shown here at work in his laboratory blasting bits of carbon with a laser. His research, along with that of Dr. Kroto, resulted in the discovery of the bucky ball—seen here as computer-generated image.**

TEACHING STRATEGY: ADVENTURE

FOCUS/MOTIVATION

Display a number of pictures showing geodesic domes built by R. Buckminster Fuller. Ask students to examine the pictures and tell what they have in common. (All are spherical and lightweight.)

CONTENT DEVELOPMENT

Discuss the Gazette article with students.
• **Why did Smalley and Kroto name their discovery of a new molecular form of carbon after Buckminster Fuller?** (The form resembled the geodesic dome architectural structures built by Fuller.)
• **How do you think Fuller might have felt about having a scientific discovery named after him?** (Students are likely to suggest that Fuller would have been honored, particularly after the existence of the fullerene was confirmed.)

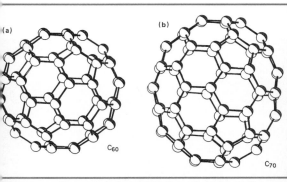

▶ A three-dimensional model of Buckminsterfullerene shows its pattern. Scientists hope that they can produce molecules that keep the same shape yet contain more carbon atoms. On the right is the first microscopic image of buckminsterfullerenes.

fellow scientists. Some said that Smalley and Kroto had failed to provide enough concrete evidence to prove their discovery. Some critics even said that the theory was too good, too simple, and too elegant to be true. But Smalley and Kroto refused to abandon their ideas. They were convinced that the Buckminsterfullerene molecule did indeed exist.

While Smalley and Kroto continued their research, other scientists around the globe began to investigate the possible existence of the bucky ball. In the fall of 1990, a good five years after Smalley and Kroto first published their findings, two major discoveries confirmed their theory. Researchers at the University of Arizona, the Max Planck Institute in Germany, and the International Business Machines Almaden Research Center in San Jose, California, produced and published the first photographs of the carbon balls. In addition, two physicists found a way to mass-produce the ball-shaped molecules. The laboratories in Arizona and Germany pioneered a method of producing the molecules by passing an electric arc between graphite electrodes. Carbon from the electrodes, which are kept in a vacuum with a small amount of helium, vaporizes and then condenses as soot. About 5 percent of this soot results in fullerene molecules.

Now scientists—even the skeptics—are excited about the fullerene's future. Because of its exceptional stability, the ball-shaped molecule has many potential applications. As a researcher at the University of Arizona explained, the fullerene may introduce a "brand new kind of chemistry." Scientists imagine that fullerenes may be able to make a new type of lubricant because the molecules roll like invisible ball bearings. The bucky balls may also be used to produce new organic compounds and in the development of better batteries. Other areas of research include the use of bucky balls on new telecommunications systems as building blocks for a new generation of high-speed computers based on light waves rather than on electricity. There is also hope of encapsulating drugs for cancer treatments in bucky balls to preserve the drugs until they reach the intended therapy site. But the busiest and most exciting area of research is in the application of fullerenes as superconductors.

So despite the fact that the bucky ball doesn't have much use on the playing fields, it undoubtedly has tremendous potential in the laboratory, as well as in the marketplace.

GAZETTE ■ 141

Additional Questions and Topic Suggestions

1. Suggest that students research the architectural career of R. Buckminster Fuller. When did he first begin building geodesic domes? Why did he turn to this type of architecture?
2. Have students investigate why the stability of a substance is important. Then, based on their findings, ask them to determine whether stability or instability increases the use of a substance and why.
3. Have students consider whether scientific discoveries are usually greeted with skepticism, and if so, why. Ask students to find out about other scientific discoveries that were not accepted at first.
4. The buckminsterfullerene was first discovered in 1985. Have students investigate whether any practical applications of the bucky ball have yet been discovered, and if so, what these applications are.

Critical Thinking Questions

1. Is the bucky ball a naturally occurring molecular form of carbon, as are graphite and diamonds? Why or why not? (Answers will vary. Many students will suggest that because the bucky ball was manufactured in the laboratory, it is not a naturally occurring substance.)
2. Why did one researcher indicate that the Buckminsterfullerene might introduce a "brand new kind of chemistry"? (Accept all logical answers.)
3. Why do you think that although there is some debate among scientists about the bucky ball, the discovery of the Buckminsterfullerene has not been discredited? (Lead students to understand that other scientists were able to produce the bucky ball.)

How does the naming of the bucky ball demonstrate that there is room for humor in scientific research? (The scientists named the bucky ball after an architect whose buildings resembled the structure of their scientific discovery, rather than after themselves or a technical or scientific term.)

The bucky ball has 32 facets, or sides.
• **Are all the sides the same?** (No, 12 sides are shaped like pentagons and 20 are hexagons.)

• **Why is the discovery of the bucky ball important?** (Because of its structure, the bucky ball is an extremely stable molecule that has many potential applications.)
• **What application might the bucky ball have to computers?** (The bucky ball might be able to be used to develop high-speed computers that operate on light waves rather than on electricity.)

INDEPENDENT PRACTICE

▶ *Activity Book*

After students have read the Science Gazette article, you may want to hand out the reading skills worksheet based on the article in the *Activity Book*.

HOW PRACTICAL IS
FLOWER POWER?

Background Information

The idea of using plant material for fuel is not new. Since ancient times, people have burned wood and other organic materials to produce energy. In the last century, however, fossil fuels such as coal, oil, and natural gas have become the leading sources of energy in industrialized nations.

The burning of fossil fuels presents many problems, and these problems are growing more severe. First, fossil fuels produce a lot of air pollutants, particularly sulfur, carbon dioxide, and nitrogen oxides. Second, fossil fuels are a nonrenewable resource. This means that once the world's supply of coal, oil, and natural gas are used up, there will be no more. Many experts believe that leading nations are consuming oil at such a rapid rate that the world could easily be out of oil before the end of the next century.

Because of the problems of air pollution and dwindling fossil fuel resources, scientists continue to search for alternative energy sources. One of these alternative sources is a group of combustible materials known as biomass. Biomass includes plants, animal wastes, and all other forms of matter that come from living things.

Wood is the main form of biomass. Other forms of combustible vegetation include corn husks, waste fibers from sugar cane, sunflowers, and seaweed. A major advantage of biomass is that it is a renewable resource.

The use of plant material for energy received considerable publicity during the gasoline crisis of the 1970s. In a search for an alternative to gasoline, scientists developed gasohol, which is a mixture of gas and ethanol. Ethanol is obtained by the action of yeast cells on various grains such as corn, wheat, and barley. Today, scientists are experimenting with cars that run entirely on ethanol.

SCIENCE GAZETTE

* HOW PRACTICAL IS *
FLOWER POWER?

Scientists are hard at work trying to find an energy substitute for oil, and it may be an ordinary green plant. But will the price be too high?

The sun rises on a summer day in the not-too-distant future. As its rays spread over the land, a farmer goes to work in his fields. His crop is almost ready for harvesting and will bring him a good price. What kind of crop is he growing? The farmer of the future is growing energy!

TEACHING STRATEGY:
ISSUE

FOCUS/MOTIVATION

Use two wooden poles and squares of green and red cloth to create a red flag and a green flag. Display the flags to the class.
• **If you saw one of these flags, what might you think it means?** (Accept all answers. Students may associate the red flag with some kind of caution or signal to stop; they may associate the flags displayed together as some kind of signal to stop or go.)

Explain to students that in the 1970s, the United States experienced a severe gasoline shortage. Because of the shortage, many filling stations ran out of gasoline early in the day or had no gasoline delivered. Long lines would form around filling stations as people tried to get gasoline. In order to inform motorists of gasoline availability, stations would display a green flag, meaning yes, or a red flag, meaning no. Sometimes a station would

This may not seem like news to you. After all, some plants, such as trees, are already sources of energy. Trees are used as firewood. And when wood is burned, energy is released in the form of heat and light. Where does this energy come from? It comes from the sun!

Each green leaf in a plant uses energy from the sun—sunlight—to build chemical molecules of sugar. In the process, the sun's energy gets locked up in the sugar molecules and other plant substances. That energy is given off when a plant or its parts are burned.

The same thing happens when coal, oil, and natural gas are burned. These fuels were formed from decaying plants and from animals that ate plants. Buried in the Earth for millions of years, the remains of the plants and animals were converted by tremendous heat and pressure into the coal, oil, and gas of today. But only a small percentage of the sun's energy originally absorbed by the plants stayed in these fuels. So the burning of fuels is not the most efficient way to get at plant energy.

FUEL FROM PLANTS

Melvin Calvin, a Nobel prize-winning chemist at the University of California at Berkeley, has worked for years to find a better way to recover the solar energy stored in plants. One way to get more energy out of plants is to take their chemicals and make them into high-energy fuels. One such fuel is alcohol. When burned, alcohol is a powerful fuel. Some high-powered race cars, for instance, run on pure alcohol.

Unfortunately, alcohol is not made directly by plants but from their sugar compounds.

▲ Substitutes for oil and gas are always being hunted. Corn, for example, is a source of alcohol, which is combined with gasoline to make the fuel gasohol. But growing and harvesting corn uses up a lot of fuel and energy—too much, maybe, to make gasohol a practical substitute for gasoline.

Sugar cane and cereal grains such as corn produce large amounts of sugars. Through a chemical process called fermentation, these sugars can be broken down into alcohol.

A mixture of gasoline and alcohol produces gasohol, which is sold as a substitute for gasoline at some filling stations. When gasohol was first marketed, it was hailed as a solution to the energy crisis. But gasohol has never really become popular or practical. The following reasons highlight why "flower power" may not become a reality in the near future.

ENERGY TO MAKE ENERGY

For one thing, a lot of energy is needed to make energy. Corn, for example, is a major source of gasohol. Yet, one scientist notes, "Corn grown in the U.S. requires a surprising amount of energy."

For example, corn plants need large amounts of fertilizer. And most of the fertilizers used for growing corn are made from

GAZETTE ■ 143

Additional Questions and Topic Suggestions

1. Discuss the energy conversions that are involved when plants are used as fuel. (Radiant energy from the sun is converted into chemical energy in plants. As plants or parts of plants are burned, the chemical energy is converted into heat energy. This heat energy may then be converted into other forms of energy, such as mechanical or electrical energy.)

2. Choose one of the following alternative energy resources and discuss its advantages and disadvantages: nuclear power, geothermal energy, wind energy, solar energy.

3. What are the advantages and disadvantages of obtaining hydrocarbon fuels directly from plants? (The main advantage is that energy-consuming processing—such as that involved in making ethanol or gasohol—is not necessary. The main disadvantage is that at the present time, scientists do not really know the amounts of hydrocarbons that can be obtained in this way, the cost of obtaining fuel from them, or the conditions that are needed for growing the types of plants that can produce hydrocarbons.)

4. Do you feel that the use of land to grow fuel-producing crops is justified when that same land could be used to produce food crops? Explain your answer. (Accept all logical answers.)

5. The making of gasohol does not really eliminate the need for oil. Why? (Gasohol is usually made from corn, and corn requires fertilizers that are petroleum products. Tractor fuel is needed for planting and harvesting the corn, and energy is needed to dry the grain and to convert the grain into alcohol. Oil is usually the source of all this energy.)

---

display an orange flag, meaning that they would sell only limited amounts of gasoline to each customer.

CONTENT DEVELOPMENT

Continue the previous discussion by explaining to students that although the gasoline crisis of the 1970s passed, energy experts are still greatly concerned about diminishing supplies of oil and other fossil fuels because fossil fuels are a nonrenewable resource.

• **What is a nonrenewable resource?** (One that cannot be replaced once it is used up.)

Emphasize that a great deal of research is being done today to develop energy sources that are renewable. One source of energy that is renewable is plants.

• **Why are plants a renewable resource?** (After plants have been used, more can always be grown.)

Point out that plants can be used as a fuel source both indirectly and directly. Fuels are made from plants indirectly by

a process that converts plant-sugar compounds into alcohol. This process is called fermentation. The alcohol can then be mixed with gasoline to produce gasohol, or it can be used to power cars that are designed to run on alcohol alone.

Obtaining fuels directly from plants is still in the experimental stages. Scientists have found that certain types of plants produce hydrocarbons that are the basic compounds in oil. Researchers are trying to find ways to extract these compounds and use them directly as fuels.

Class Debate

Have students work in small groups. Ask each group to write a proposal for a research project that would develop a plant energy source. Included in the proposal should be persuasive arguments in favor of the project. Then have each group present their proposal to the class. Members of the class should challenge the group and debate the pros and cons of the proposal. Finally, have students vote to determine whether each proposal is accepted or rejected.

▲ **Melvin Calvin stands in a field of gopher plants at his ranch in California. Calvin and other scientists like him hope to use gopher plants, and other plants, as new sources of fuel. Will gopher plants help to replace oil wells? Some scientists think so.**

oil. And, of course, it takes a lot of energy to get oil out of the ground and to move it to where it is used. Other ways that energy is used to produce corn are the burning of tractor fuel for planting, cultivating, and harvesting; the production of heat for grain drying; and, if irrigation is needed, energy to pump water.

Most of this energy comes from oil or natural gas. Moreover, converting the grain to alcohol requires even more energy. Equally important, if land is used to produce fuel, is how the world's food supplies would be affected. In other words, does it make sense to "farm" fuel? It makes sense only if no one goes without food and the fuel is inexpensive. So far, this does not seem to be the case.

However, scientists have come up with a possible way to get around the problem—

GAZETTE

144 ■ GAZETTE

farm plants that *directl* produce fuel. Oil from plants sounds impossible doesn't it? Some plant however, produce hydr carbons, which are th basic compounds in oi Among these plants are th rubber tree and its rela tives. Some of these plant resemble cacti. The hydro carbons are in the milky sa of the plants. They might b used to produce inexpensiv fuels. Melvin Calvin an other researchers have bee trying to find a way to pre duce fuel from plants.

Calvin has also done re search on the gopher plant which grows in the wester part of the United State He and scientists working with him are grow ing gopher plants on experimental plots o Calvin's own ranch in California. They hop to determine the amount of hydrocarbon the gopher plants produce and the cost o making fuel from them.

Because gopher plants are wild and hav not been cultivated before, little is know about methods of growing them and the con ditions they require. The sap of the gophe plant, moreover, cannot be tapped as easi ly as that of some of its relatives.

Perhaps the gopher plant will prove to b too difficult to grow. Or possibly the amoun of hydrocarbons that can be produced from it will not be large enough to make it a goo fuel source. On the other hand, the gophe plant could turn out to be a substitute fo an oil well. And if it is not, perhaps som other plant will be.

There are, of course, other sources of in expensive energy we should look to: sola energy, nuclear energy, natural gas energy old-fashioned coal, oil, and wind energy. Bu each has its problems. Can "flower power compete with these energy sources? What d you think?

ISSUE (continued)

REINFORCEMENT/RETEACHING

Review with students the many "hidden costs" involved in producing fuel from plants. These include planting, cultivating, and harvesting the crops; fertilizing the crops; providing irrigation where necessary; converting grain to alcohol; extracting hydrocarbons from plants such as the gopher plant; and converting plant hydrocarbons into fuel.

GUIDED PRACTICE

Skills Development

Skill: Expressing an opinion

Have students consider the following statement and then decide whether they agree or disagree and why:

The main reason for developing fuels from plants is not to save money but to save the world's oil supply. Even if plant fuels are expensive, they are renewable—and this makes them a bargain in the long run.

INDEPENDENT PRACTICE

▶ *Activity Book*

After students have read the Science Gazette article, you may want to hand out the reading skills worksheet based on the article in the *Activity Book*.

SCIENCE GAZETTE:
The Right Stuff—
PLASTICS

Cars, buildings, and appliances of the twenty-first century will be lighter, safer, and stronger than anything we have today.

Standing apart from a group of her classmates, Jennifer bounced her glass ball on the sidewalk. She was not paying attention to the other members of her class or to what her teacher, Ms. Parker, was saying.

As the sparkling ball bounced, Jennifer was barely aware of the sidewalk changing color to reduce the glare of sunlight. She did not notice a young woman hurrying along the street carrying an auto engine in one hand and groceries in the other. She did not look up as cars sped by towing boats and buses by plastic threads.

Jennifer and her Design for Living class were on a field trip to a brand-new "house of the future." The house had just been completed that year: 2093. It certainly made the students' homes seem old-fashioned.

For days before the field trip, Ms. Parker had talked about the house of the future. She had described how safety, security, cleaning, maintenance, home education, and communications systems in the house were completely controlled by a central computer. Ms. Parker had shown the students how the combination of a waste-recycling unit, solar cells, and wind turbines met all of the house's energy needs. Ms. Parker had spent a long time talking about the new materials that made such a house

GAZETTE ■ 145

FUTURES IN SCIENCE

THE RIGHT STUFF— PLASTICS

Background Information

This fanciful article depicts life in the year 2093—just about one hundred years from now. As a result of a "materials revolution," familiar objects such as cars, refrigerators, and furniture are made of strong, flexible, lightweight plastic instead of heavy, rigid materials such as metal and wood.

The article stresses that the plastics of the future are being developed today, in the 1990s. The plastics are being made from a new family of polymers that are silicon atoms instead of carbon atoms. Described in the article are applications of these materials that may come about within the next hundred years.

In addition to plastics, the article touches on some other materials and inventions that may be part of life in the late twenty-first century. These include the use of solar and wind energy for household power; laser brake systems in automobiles; homes controlled by central computer systems; elastic glass; and building materials that change color to adapt to sun glare.

TEACHING STRATEGY: FUTURE

FOCUS/MOTIVATION

Begin by asking students the following questions:
• **What do you think life would be like without plastic?** (Accept all answers.)
• **How many items have you used so far today that are made of plastic?** (Answers will vary.)
• **How many items can you think of in your home that are made of plastic?** (Answers will vary.)
• **What types of materials do you think were used to make these items before plastics were invented?** (Accept all answers.)

CONTENT DEVELOPMENT

Use the previous discussion to introduce the idea that even though we tend to take plastics for granted, they are a rather recent invention. Plastics are made of polymers, and the first polymer was manufactured in 1909. Items that are today made of plastic were once made of materials such as wood, glass, china, and metal. Other synthetic materials that have been developed during the twentieth century include synthetic rubber and synthetic fabrics, such as nylon and polyester.

Once students recognize how much familiar appliances and other items have changed during the last hundred years, they may find it easier to imagine how much these same items might change during the next hundred years.

Additional Questions and Topic Suggestions

1. Have you ever felt that you like old-fashioned things better than modern things or that you prefer things made of natural materials rather than synthetic materials? If so, give examples and explain why you feel the way you do.

2. Based on the article, do you think you would enjoy living in the year 2093? Why or why not? (Accept all answers.)

3. Imagine that you have been asked by a museum to create a "time capsule." The time capsule will be preserved to show future generations what life is like in the United States in the late twentieth century. Decide what items from your home, school, or personal possessions you would include to give an accurate picture of twentieth-century materials and technology.

possible. Several times, she had drawn pictures of organic and inorganic molecules on the classroom computer screen.

Even with all this preparation, Jennifer did not care about the house. She was interested only in the designs of the past. Jennifer's real love was for twentieth-century antiques. In her opinion, the appliances, furnishings, and materials of 2093 were boring and ugly.

"Jennifer!" Ms. Parker's voice cut through the group.

Jennifer scooped up the bouncing ball and thrust it into her shoulder bag.

"Yes, Ms. Parker?"

"Please come forward and give us the benefit of your views on home design."

As Jennifer moved to the front of the group, she knew she was on the spot. The class was already aware of Jennifer's "good-old-days" views, which contrasted with Ms. Parker's enthusiasm for modern materials and designs.

Nervously, Jennifer started to talk. "Plastics, silicones, polymers...I don't think the scientists of the late twentieth century did us a favor by developing all this phony stuff. I wish we could go back to using wood, steel, aluminum, copper, and real glass. Remember those beautiful steel-and-chrome cars and the polished wood furniture in the design museum?"

"Yes," Ms. Parker agreed, "many things produced in the last century were beautiful and well made. It's easy to admire them in museums. But I don't think we'd find them convenient to live with every day."

"I would," Jennifer insisted.

"All right," responded Ms. Parker, "let's hop into a mental time machine and find out. What age should we go back to?"

"How about the 1990's?" Jeremy suggested.

"Perfect," said Ms. Parker. "The 1990's were just the beginning of the great materials revolution. Great changes in industry, electronics, and materials development began at that time. By the way, Jennifer, you'd better leave your bouncing ball behind. They didn't have elastic glass in the 1990s. Light-

▲ Plastics are now being used to desig[n] technology of the future, such as this mod[el] of a refrigerator-freezer.

sensitive building materials that change co[lor] or to reduce glare were also unheard o[f.] However, the research to produce these m[a]terials was already under way."

A TRIP BACK IN TIME

The house of the future was forgotten [as] Ms. Parker and the class moved to a nea[r]by park to talk about materials of the pas[t.]

Ms. Parker began by asking the student[s,] "What would be the first thing you'd noti[ce] in the world of the 1990s?"

"Heaviness. The great weight of almo[st] everything," Jeremy volunteered.

"Go on," Ms. Parker said.

"Appliances such as refrigerators, stove[s,] washing machines, and air conditioners we[re] still made of metal at that time. They we[re] so heavy that it was almost impossible f[or] one person to lift any of them. And a lot [of] everyday things were much heavier than the[y] are now. Many food items in supermarke[ts] still came in metal cans and glass bottles. [A] bagful of those containers could weigh a l[ot.] It wasn't until around 2000 that lightweigh[t,] tough plastic containers had completely r[e]placed them."

FUTURE (continued)

GUIDED PRACTICE

Skills Development

Skill: Relating concepts

Have students think of examples of machines, appliances, and other familiar items that have changed in appearance and construction during the last century. Some of these items include automobiles, vacuum cleaners, radios, typewriters, com-

puters, sewing machines, bicycles, and record players. Students may enjoy collecting pictures of these items as they have looked at different times in history and then comparing them with modern versions.

Discuss with students some reasons for the changes they see. Explain that items such as automobiles and record players have changed largely as a result of improved technology. Items such as vacuum cleaners have changed primarily because of new materials. Some old-fashioned vac-

uum cleaners were made of heavy metal, whereas most of today's vacuum cleaner[s] are made of lightweight plastic. Comput[ers] have changed drastically because o[f] both improved technology and new ma[terials. The first computers were a mass o[f] metal cranks and gears and were so mas[sive that they took up a whole room. To[day's computers are so small and lightweight that they can be picked up and carried as easily as a briefcase.

SAFER FUTURE

Turning to another student, Ms. Parker asked: "Carlee, what would you notice most about life in the 1990s?"

"That it wasn't safe," Carlee responded.

"Why?"

Carlee thought for a moment and then said, "I guess what I was thinking of was the danger of riding in a 1990 car. I've seen pictures of how those old metal cars hurt people in accidents. Sometimes the heavy metal engines and batteries in the front end were pushed back to where people were sitting. Sharp pieces of broken metal and glass were all over the scene of an accident. Cars could blow up or catch fire."

"That can't happen now," Marian said. "Our cars, including the engines, are made of super plastics and silicones inside and out. These materials are very light and strong, and they bounce. Even if a laser brake system fails, no one can get seriously hurt."

"Cars not only are lighter and safer now," Jeremy added, "they require less energy to run. The changeover from metal to plastic engines led to great energy savings. And the changeover since then to solar battery-powered cars has meant even greater savings. If we'd kept on using heavy metal cars, the world might have run out of oil and other materials by now."

"You're quite right, Jeremy," Ms. Parker agreed. "Let's sum up what's happened since the 1990s. A great materials revolution started at that time. Chemists discovered how to produce polymers: very long chains and loops of carbon, oxygen, hydrogen, and nitrogen atoms.

"Around the same time, other scientists created a new family of polymers. Silicon atoms were used instead of carbon atoms, so the materials were called silicones. The result of all this chemistry was a new range of super-strong light, cheap plastics.

"The new materials can be made into anything that one made with metal, wood, glass, or ceramic. In fact, we can do many things with combinations of the new materials that we couldn't do with the old materials, such as making a transparent bouncing ball." Ms. Parker smiled at Jennifer.

Jennifer smiled back, still not convinced of the advantages of living in 2093.

▲ Cars of the future may be made entirely of plastics and powered by solar batteries. Such a car would be lightweight, durable, and clean and inexpensive to operate.

Critical Thinking Questions

1. Do you think that the lightweight plastic cars of 2093 would be totally safe, as the article describes? Support your answer. (Accept all logical answers. Students may wish to consider the possibility that a powerful impact at high speed might still be dangerous, even with light, strong materials that bounce.)

2. Why do you think a plastic engine would be especially useful in a racing car? (Answers may vary. A logical answer is that the lighter an object is, the less force and therefore less energy it takes to accelerate the object. Because the primary purpose of a racing car is speed, a lightweight plastic engine would be extremely practical.)

3. What aspect of the article indicates that some of the modern inventions described are likely to become a reality in the near future? (The article indicates that the materials for these items began to be produced in the 1990s. This means that many of the materials for the items have already been invented.)

REINFORCEMENT/RETEACHING

Have students list the items described in the article that are clearly of the future. These items include Jennifer's ball made of elastic glass; an auto engine that can be carried in one hand; plastic threads that can tow boats and cars; a computer-controlled house; solar cells and wind turbines, used for household power; and a sidewalk that changes color to reduce the glare of the sun.

INDEPENDENT PRACTICE

▶ *Activity Book*

After students have read the Science Gazette article, you may want to hand out the reading skills worksheet based on the article in the *Activity Book.*

For Further Reading

If you have been intrigued by the concepts examined in this textbook, you may also be interested in the ways fellow thinkers—novelists, poets, essayists, as well as scientists—have imaginatively explored the same ideas.

Chapter 1: Atoms and Bonding

Bond, Nancy. *The Voyage Begun*. New York: Macmillan.

Botting, Douglas. *The Giant Airships*. New York: Time-Life Books.

Bracy, Norma M. *Salt*. Atwater, CA: Book Binder.

Henry, O. "The Diamond of Kali." In *The Complete Works of O. Henry*. New York: Doubleday.

Lee, Martin. *Paul Revere*. New York: Watts.

Maupassant, Guy de. "The Diamond Necklace." In *The Best Stories of Guy de Maupassant*. New York: Airmont.

Spinelli, Jerry. *Space Station Seventh Grade*. Boston: Little, Brown.

Chapter 2: Chemical Reactions

Anderson, Madelyn K. *Environmental Diseases*. New York: Watts.

Boning, Richard A. *Horror Overhead*. Baldwin, NY: B. Loft.

Burchard, S. H. *The Statue of Liberty: Birth to Rebirth*. New York: Harbrace Junior Books.

Fox, Mary V. *Women Astronauts: Aboard the Space Shuttle*. Englewood Cliffs, NJ: Messner.

Krementz, Jill. *The Fun of Cooking*. New York: Knopf.

Le Guin, Ursula K. *A Wizard of Earthsea*. New York: Bantam.

Stewart, Gail. *Acid Rain*. San Diego, CA: Lucent.

Tchudi, Stephen. *Soda Poppery: The History of Soft Drinks in America*. New York: Macmillan.

Chapter 3: Families of Chemical Compounds

Carter, Alden R. *Up Country*. New York: Putnam.

Dolan, Edward F., Jr. *Great Mysteries of the Ice and Snow*. New York: Putnam.

Fox, Paula. *The Moonlight Man*. New York: Bradbury.

Hendershot, Judith. *In Coal Country*. New York: Knopf.

Kirschmann, John, and Lavon Dunne. *Nutrition Almanac*. New York: McGraw-Hill.

Smith, Robert Kimmel. *Jelly Belly*. New York: Delacorte.

Chapter 4: Petrochemical Technology

Dineen, Jacqueline. *Wool*. Hillside, NJ: Enslow.

Eliot, George. *Silas Marner*. New York: New American Library.

Ferber, Edna. *Giant*. New York: Fawcett.

Merrill, Jean. *The Pushcart War*. New York: Addison Wesley.

Tarkington, Booth. *The Magnificent Ambersons*. Bloomington, IN: Indiana University Press.

Trier, Mike. *Super Car*. New York: Watts.

Whyman, Kathryn. *Plastics*. New York: Watts.

Chapter 5: Radioactive Elements

Careme, Maurice. *The Peace*. San Marcos, CA: Green Tiger.

Dank, Milton. *Albert Einstein*. New York: Watts.

Hersey, John. *Hiroshima*. New York: Bantam.

Keller, Mollie. *Marie Curie*. New York: Watts.

Mikliowitz, Gloria D. *After the Bomb*. New York: Scholastic.

Nardo, Don. *Chernobyl*. San Diego, CA: Lucent.

$\mathbf{A}$ctivity $\mathbf{B}$ank

Welcome to the Activity Bank! This is an exciting and enjoyable part of your science textbook. By using the Activity Bank you will have the chance to make a variety of interesting and different observations about science. The best thing about the Activity Bank is that you and your classmates will become the detectives, and as with any investigation you will have to sort through information to find the truth. There will be many twists and turns along the way, some surprises and disappointments too. So always remember to keep an open mind, ask lots of questions, and have fun learning about science.

Activity Bank

COOPERATIVE LEARNING

Hands-on science activities, such as the ones in the Activity Bank, lend themselves well to cooperative learning techniques. The first step in setting up activities for cooperative learning is to divide the class into small groups of about 4 to 6 students. Next, assign roles to each member of the group. Possible roles include Principal Investigator, Materials Manager, Recorder/Reporter, Maintenance Director. The Principal Investigator directs all operations associated with the group activity, including checking the assignment, giving instructions to the group, making sure that the proper procedure is being followed, performing or delegating the steps of the activity, and asking questions of the teacher on behalf of the group. The Materials Manager obtains and dispenses all materials and equipment and is the only member of the group allowed to move around the classroom without special permission during the activity. The Recorder, or Reporter, collects information, certifies and records results, and reports results to the class. The Maintenance Director is responsible for cleanup and has the authority to assign other members of the group to assist. The Maintenance Director is also in charge of group safety.

For more information about specific roles and cooperative learning in general, refer to the article "Cooperative Learning and Science—The Perfect Match" on pages 70–75 in the *Teacher's Desk Reference.*

ESL/LEP STRATEGY

Activities such as the ones in the Activity Bank can be extremely helpful in teaching science concepts to LEP students—the direct observation of scientific phenomena and the deliberate manipulation of variables can transcend language barriers.

Some strategies for helping LEP students as they develop their English-language skills are listed below. Your school's English-to-Speakers-of-Other-Languages (ESOL) teacher will probably be able to make other concrete suggestions to fit the specific needs of the LEP students in your classroom.

• Assign a "buddy" who is proficient in English to each LEP student. The buddy need not be able to speak the LEP student's native language, but such ability can be helpful. (**Note:** *Instruct multilingual buddies to use the native language only when necessary, such as defining difficult terms or concepts. Students learn English, as all other languages, by using it.*) The buddy's job is to provide encouragement and assistance to the LEP student. Select buddies on the basis of personality as well as proficiency in science and English. If possible, match buddies and LEP students so that the LEP students can help their buddies in another academic area, such as math.

• If possible, do not put LEP students of the same nationality in a cooperative learning group.

• Have artistic students draw diagrams of each step of an activity for the LEP students.

You can read more about teaching science to LEP students in the article "Creating a Positive Learning Environment for Students with Limited English Proficiency," which is found on pages 86–87 in the *Teacher's Desk Reference.*

Activity Bank

UP IN SMOKE

BEFORE THE ACTIVITY

At least one day prior to the activity, gather the required materials for each group.

PRE-ACTIVITY DISCUSSION

Write the formulas for the seven compounds on the chalkboard. Then focus student attention on the purposes of this activity by asking the following questions.
• **What do these seven compounds have in common?** (They all contain the symbol Cl, which is the symbol for the element chlorine.)
• **What is the variable in each formula?** (The metal with which chlorine is bonded.)

SAFETY TIPS

Remind students about the rules to be followed when working with a Bunsen burner. Make sure students are wearing safety goggles. Caution them on the use of any acid, even weak hydrochloric acid. Remind them about proper disposal of acids.

TEACHING STRATEGY

Circulate among student groups to be sure they are using the Bunsen burner correctly. Also make sure they clean their nichrome or platinum wire prior to each test.

DISCOVERY STRATEGIES

Discuss how the activity relates to the chapter by asking questions similar to the following.
• **Why were the different metals being tested bonded to chlorine?** (Chlorine is a very active element that combines readily with many metals. By keeping chlorine the control, the metal became the variable and could thus be tested and observed.)
• **Why is a flame test a qualitative test?** (A qualitative test indicates the presence of a particular substance. In this case, the test identifies specific metals.)

WHAT YOU SEE

LiCl: crimson; CaCl$_2$: yellow-red; KCl: violet; CuCl$_2$: blue-green; SrCl$_2$: red; NaCl: yellow; BaCl$_2$: green-yellow.

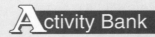
Electrons surrounding an atomic nucleus are usually located in particular energy levels. Sometimes, however, an electron can absorb extra energy, which forces it up into a higher energy level. An electron in a higher energy level is unstable. Eventually the electron will fall back to its original position, and as it does so, it releases its extra energy in the form of light.

Different elements can absorb and release only certain amounts of energy. The amount of energy determines the color of the light that is given off. Thus the color of light given off can be used to identify particular elements. In this activity you will give extra energy in the form of heat to different elements and observe the colors given off.

Materials You Need

nichrome or platinum wire
cork
Bunsen burner
hydrochloric acid (dilute)
distilled water
7 test tubes
test-tube rack
7 chloride test solutions

Procedure

1. Label each of the test tubes with one of the following compounds: LiCl, CaCl$_2$, KCl, CuCl$_2$, SrCl$_2$, NaCl, BaCl$_2$. Pour 5 mL of each test solution into the correctly labeled test tube.
2. Push one end of the wire into the cork and bend the other end into a tiny loop.
3. Put on your safety goggles. Hold the cork and clean the wire by dipping it into the dilute hydrochloric acid and then into distilled water. Then heat the wire in the blue flame of the Bunsen burner until the wire glows and you can no longer see colors in the flame.
4. Dip the wire into the first test solution. Place the end of the wire at the tip of the inner cone of the burner flame. Record the color given to the flame in a data table similar to the one shown.
5. Clean the wire by repeating step 3. Repeat step 4 for the other 6 test solutions. Make sure you clean the wire after each test.

What You See

DATA TABLE

Compound	Color of Flame
LiCl	
CaCl$_2$	
KCl	
CuCl$_2$	
SrCl$_2$	
NaCl	
BaCl$_2$	

What You Can Conclude

1. Why do you think all of the compounds you tested were bonded to the same element—chlorine?
2. Why did you have to clean the wire before each test?
3. How do your observations compare with those of your classmates?
4. How can you use the flame test to identify a certain element?

WHAT YOU CAN CONCLUDE

1. All the compounds must be bonded to the same element so that one can be sure that the differences in the observed colors are due to differences in the metals. Otherwise, one would not be able to identify the metal by the characteristic color it produced.
2. Some of the solution from the previous test might remain on the wire. The color seen would then be a combination of the two metals. It would not be possible to isolate one from the other.

3. All students should observe the same colors.
4. As observed from this activity, certain elements give off characteristic colors. Given an unknown solution, a flame test could be performed to find out what color is given off. A comparison of that color with the characteristic colors recorded in the data table would identify the metal in the unknown solution.

HOT STUFF

If you look around, you will find that many of the objects and structures you depend on every day are made of metal. Metals have quite interesting and useful characteristics. One of the most important of these is the fact that metals are excellent conductors of both heat and electricity. In this activity you will find out just how well metals conduct heat.

For this activity, you will need several utensils (spoons or forks) made of different materials, such as silver, stainless steel, plastic, wood, and so forth. You will also need a beaker (or drinking glass), hot water, a pat of frozen butter, and several small objects (beads, frozen peas, popcorn kernels, or raisins).

Press a small glob of butter onto the top of each utensil. Make sure that when the utensils are stood on end, the butter is placed at the same height on each. Be careful not to melt the butter as you work with it. Press a bead, or whatever small object you choose, into the butter. Stand the utensils up in the beaker (leaning on the edge) so that they do not touch each other. Pour hot water into the beaker until it is about 6 cm below the globs of butter.

Watch the utensils for the next several minutes. What do you see happening?

Make a chart listing the material each utensil is made of, and the order in which the bead fell into the water.

Which do you expect to fall first? Which actually does?

Combine your results with those of your classmates. Make a class chart showing all of the materials used and the order in which the beads fell.

Activity Bank

HOT STUFF

BEFORE THE ACTIVITY

1. Gather all materials at least one day in advance. You should have enough materials to meet your class needs, assuming groups of four to six students.

2. Make sure you have a variety of utensils so that students will be able to see the difference in conductivity among the substances. If you cannot find wooden spoons or forks, you might consider using chopsticks.

3. Make sure you keep the butter frozen until it is to be used.

PRE-ACTIVITY DISCUSSION

Focus student attention on the purposes of this activity by asking the following questions.

• **What is conductivity?** (The ability to conduct heat or electricity.)

• **What determines the conductivity of a** substance? (The bonding that occurs between atoms.)

• **Do all substances have the same conductivity?** (No. Because conductivity is determined by atomic bonding, different substances will conduct differently.)

• **Do you know what name is given to substances that do not conduct heat or electricity?** (Insulators.)

TEACHING STRATEGY

1. Circulate among the student groups to be sure they have set up their materials correctly. Be sure that the utensils are not touching each other.

2. Check to be sure that students have placed the glob of butter at the same height on each utensil. Discuss why this placement is so important to the results of the activity.

DISCOVERY STRATEGIES

Discuss how this activity relates to the concepts developed in the chapter by asking questions similar to the following.

• **Which material do you think will be the best conductor? The poorest?** (Accept all student answers and as you do, discuss the merits of each response and the reasoning behind it.)

• **Why do you think butter was used?** (Accept all logical answers. What you should elicit from students is that butter melts rather easily and thus observations can be made in a fairly short time.)

• **Why must the glob of butter be at the same height on each utensil?** (Conductivity is the tested property. If height varies, one material will be favored over the others and results will not be conclusive.)

• **How do the actual results compare with your hypothesis?** (Answers will vary.) **With the results of your classmates?** (Answers will vary. However, if all groups did the activity correctly, the results should be the same.)

• **How do the results of this activity relate to your own experiences?** (Accept all logical answers and discuss each case as students report them.)

ANSWERS

■ The butter begins to melt and slide down the utensil.

■ The butter on the metal utensil will melt first, causing the bead to slide into the water. This is because the metal is the best conductor of the heat from the water.

Activity Bank

BEFORE THE ACTIVITY

Gather all materials at least one day in advance. Plan on having students work in groups of four to six.

PRE-ACTIVITY DISCUSSION

Focus student attention on the purposes of this activity by asking the following questions.

• **Why do you think certain substances mix and others do not?** (Accept all logical answers. What you should be looking for here is an awareness of the role bonding plays in substances mixing, or dissolving. Generally, like dissolves like, where like refers to the type of bond: ionic or covalent.)

• **What might be true about water that it dissolves almost everything?** (Water has properties that make it dissolve both ionic and covalent substances. Actually, water is a polar covalent molecule.)

TEACHING STRATEGY

1. Make sure students have scattered the drops of food coloring on the milk. There should be sufficient room between drops for students to observe the movement of the drops when the detergent is added. Also make sure students do not add detergent to every food-coloring drop.

2. Suggest to students that the detergent used might play a role in the results. Have them repeat the activity in such a way that this hypothesis can be tested.

DISCOVERY STRATEGIES

Discuss how the activity relates to the chapter by asking questions similar to the following.

• **Would the type of detergent used alter the results?** (If it did, it would only be in the speed with which and the extent to which it spread the food coloring.)

• **How do the results of this activity apply to actions in everyday life?** (Accept all logical answers. The cleansing action of soaps and detergents is the answer you wish to elicit. This cleansing action is the same for all soaps and detergents. A detergent molecule has an end that dissolves readily in water. This is its polar end. The

Have you ever squeezed a drop of dishwashing detergent into a pot full of greasy water to watch the grease spread apart? The reason this happens has to do with the molecular structure of both the grease and the detergent. Rather than bonding together, these molecules (or at least parts of them) repel each other and move away. In this activity, you will experiment with a similar example of substances that rearrange themselves when mixed together.

What You Will Need

baking sheet or roasting pan
milk (enough to cover the bottom of the sheet or pan)
food coloring of several different colors
dishwashing detergent

What You Will Do

1. Pour the milk into the baking sheet or pan until the bottom is completely covered.

2. Sprinkle several drops of each different food coloring on the milk. Scatter the drops so that you have drops of different colors all over the milk.

3. Add a few drops of detergent to the middle of the largest blobs of color. What do you see happening?

■ Can you propose a hypothesis to explain your observations?

other end of the detergent molecule does not dissolve in water. This is its nonpolar end. This nonpolar end, however, does dissolve in fats, grease, etc. With some agitation followed by rinsing, the water molecules will carry away the detergent molecules, the nonpolar ends of which are dissolved in the grease, fat, etc. Thus the "dirt" is carried away.)

WHAT YOU WILL DO

3. The colors spread out across the surface of the milk.

■ Because the detergent spreads out, taking the color with it, it would seem that the detergent is rearranging itself so that part of its structure has the least contact possible with the milk. The reason is that part of each detergent molecule is hydrophobic and moves away from the water in the milk. The detergent spreads itself over the surface so that the hydrophobic part of each molecule can stick up into the air.

POCKETFUL OF POSIES

Can you picture a meadow filled with wild flowers ranging through all the colors of the rainbow? The beautiful colors of flowers depend on combinations of chemicals carefully selected by Nature. But just as they are formed, they can also be destroyed. In this activity you will create a chemical reaction that affects flower colors.

You will need several flowers of different kinds and colors, a large jar or bottle with a lid (a clean mayonnaise jar or juice bottle will do and a plastic lid is preferable), a rubber band, scissors, and household ammonia (about 50 mL).

Procedure

1. Gather the flowers so that all of the stems are in a bunch. Use the rubber band to hold the stems together. You may have to twist it around the stems more than once.

2. Cut a large hole in the jar lid with the scissors. The gathered stems of the flowers must be able to fit snugly through the hole. **CAUTION:** *Be careful when using sharp instruments.*

3. Push the stems through the hole so that when the lid is placed on the jar, the flowers will be suspended inside the jar.

4. Pour a little ammonia into the jar—enough to cover the bottom. **CAUTION:** *Do not breathe in the ammonia vapors.*

5. Carefully place the lid with the flowers on the jar. Look at the flowers after 20 to 30 minutes. Do you observe any changes in them? If so, what do you see happening?

6. Compare your results with those of your classmates who may have used different flowers. Record the overall results.

Thinking It Through

- The pigments that give flowers their beautiful colors are present along with chlorophyll, which is green. Chlorophyll is the substance that makes photosynthesis (the food-making process in plants) possible. What do you think happened during this activity to explain your observations?

- In tree leaves, colorful pigments are also present along with chlorophyll, but in this case the green chlorophyll hides the colorful pigments. What must happen to give leaves their stunning autumn colors?

SAFETY TIPS

Warn students to avoid breathing in the ammonia vapors.

TEACHING STRATEGY

1. You can decrease the number of flowers in each bunch if you want to break the class into several groups. You can also give different colored flowers to different groups.

2. The hole in the lid of the jar must be just large enough for the stems of the flowers to fit through. Show students that the stems should fit through snugly. If the hole is too large, the flowers will fall through.

DISCOVERY STRATEGIES

Discuss how the activity relates to the chapter ideas by asking questions similar to the following.

- **What happened to the flowers during this activity?** (They were bleached.)
- **Can the flowers be returned to their original condition?** (The flowers have been permanently changed.)
- **How do you know that a chemical reaction has occurred?** (The original substances have in some way changed into different substances. In particular, the colored pigments in the flowers have been altered.)

PROCEDURE

5. Most of the colors disappear. Red, pink, and purple flowers turn green. White and yellow flowers stay the same.

6. Red, pink, and purple flowers will turn green. White and yellow flowers do not change.

THINKING IT THROUGH

- Students should realize that the ammonia was involved in a chemical reaction that destroyed the colorful pigments in some of the flowers and left the chlorophyll. Actually, it is the ammonia gas that rises from the liquid that destroys some of the colors but not others. When colors that hide the green chlorophyll are destroyed, the green color shows.

- The chlorophyll in the leaves must either be destroyed or its production must be stopped. When the chlorophyll color no longer overpowers the colored pigments, the colors of the leaves can be seen.

Activity Bank

POCKETFUL OF POSIES

BEFORE THE ACTIVITY

Gather the materials at least one day prior to the activity. Make sure you have enough materials for each group in the class.

PRE-ACTIVITY DISCUSSION

Review the topic of chemical reactions with students. Then you may want to begin a discussion by asking questions similar to the following.

- **Are you familiar with bleach and what it is used for?** (Students should be familiar with bleach. They may suggest such applications as cleaning clothes or lightening denim clothes.)

- **Do you have any clothes that have been faded by bleach?** (Answers will vary. Students may suggest clothes that were purposely bleached as well as clothes that have been ruined by accident.)

Activity Bank

POPCORN HOP

BEFORE THE ACTIVITY

Gather the materials at least one day prior to the activity. Make sure you have enough materials for each group in the class. You can assign different materials—popcorn kernels, raisins, or mothballs—to different groups if you like.

PRE-ACTIVITY DISCUSSION

You may wish to review the type of chemical reaction that occurs between vinegar and sodium bicarbonate. A discussion can then be started by asking questions similar to the following.

• **Do you know what the purpose of an antacid tablet is?** (Most students will know that antacids are used to relieve stomachaches.)

• **Have you ever seen an antacid tablet added to water? Describe what happened.** (Students will probably know that the tablet fizzes and quickly fills the water with bubbles.)

• **Do you know what is happening when an antacid tablet is added to water?** (Lead students to realize that the tablet is broken down in a chemical reaction that releases carbon dioxide bubbles.)

TEACHING STRATEGY

1. The food coloring is simply to make it easier to see the kernels as they rise and fall. It is not necessary to make the water dark in color.

2. Advise students not to add too many popcorn kernels because they might bunch up in the bottom of the container and become too heavy to be lifted by the bubbles.

DISCOVERY STRATEGIES

Discuss how the activity relates to the chapter ideas by asking questions similar to the following.

• **How do you know that a chemical reaction took place?** (The production of bubbles that did not exist before the vinegar and sodium bicarbonate were mixed together indicates that a chemical reaction took place. Both substances have changed as a result of the reaction.)

• **How can you relate this activity to the**

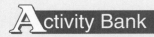

What do you see when you pour soda or other carbonated drinks into a glass? From experience, you probably know that you see bubbles continually rising to the top—thanks to the carbonation. In this activity you will create a chemical reaction similar to the one occurring in soda and you will use it to make popcorn kernels hop!

Materials

large, clear drinking glass, beaker, or jar
15 mL (1 Tbsp) baking soda
food coloring (a few drops)
45 mL (3 Tbsp) vinegar
popcorn kernels or raisins or mothballs (handful)
stirrer (or long cooking utensil)

Procedure

1. Fill the glass container with water.

2. Add about 15 mL of baking soda, a few drops of food coloring, and stir well.

3. Drop in the popcorn kernels (or raisins/mothballs) and stir in 45 mL of vinegar. Watch the kernels for the next several minutes. What do you observe happening? (If the action slows down, add more baking soda.) Explain what you see in terms of chemical reactions.

■ In an effort to preserve the natural environment, people are beginning to use Earth-friendly cleaning products. For example, rather than dumping poisonous chemicals into a sink, a mixture of hot water, vinegar, and baking soda can be used to clean drains. Why do you think this works?

bubbles you find in soda? (Lead students to realize that the bubbles in soda rise as do the bubbles in this activity. The carbonation is forced into soda under pressure and is released when the soda's container is opened.)

PROCEDURE

3. The popcorn kernels rise to the top of the water and then drop back down in a repeating process. The reason is because the baking soda and vinegar react to form bubbles of carbon dioxide. The bubbles

attach themselves to the kernels and carry them to the surface. When the bubbles burst at the surface, the kernels sink.

■ The baking soda and the vinegar react to form bubbles, as seen in the activity. The movement of the bubbles can lift dirt and grime off the sides of the drain pipe. When flushed with water, the dirt will be cleaned away without destroying the environment.

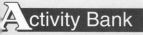

TOASTING TO GOOD HEALTH

Have you ever been given toast when you weren't feeling well? For some reason, toasted bread seems easier on your digestive system than untoasted bread does. In actuality, the reason is no mystery. It has to do with a chemical reaction involving the heat from your toaster. In this activity you will discover the difference between plain bread and toasted bread.

Materials

slice of white bread
slice of white toasted bread
household iodine (5 mL or 1 tsp)
drinking glass or 250-mL beaker
baking dish (or bowl with a flat bottom)
spoon (measuring spoon would be helpful)

Procedure

1. Fill the drinking glass or beaker half-full with water.
2. Mix 5 mL (about 1 tsp) of iodine into the water. Carefully pour the water-iodine solution into the baking dish.
3. Tear off a strip (about 2 cm wide) from the plain slice of bread. Dip the strip in the solution.
   ■ Do you observe any changes in the bread?
4. Tear off a strip of the same size from the toasted bread. Dip this strip in the solution.

■ Do you observe any changes in the toast?

■ When starch and iodine are combined, they react to form starch iodide, which is a bluish-purple. For this reason, iodine is used to test for starch. Knowing this, what can you learn from your observations?

■ What type of chemical reaction is involved in toasting—endothermic or exothermic?

■ The process of food digestion begins in your mouth. Part of this process involves breaking starches down into simpler substances. As a result of doing this activity, can you now explain why toast is sometimes recommended when you are not feeling well?

---

made of a long chain of sugars. The chain must be broken down before it can be fully digested.)
• **Can you list some foods that are made up of starches?** (Students should list potatoes, pasta, and bread, for example.)

TEACHING STRATEGY

If the toasted strip is not toasted through to the center, that part of the bread will respond the same way to the solution as the untoasted bread. Students should not think that this is a mistake.

DISCOVERY STRATEGIES

Discuss how the activity relates to the chapter ideas by asking questions similar to the following.
• **Is bread a starch?** (Students should now realized that bread belongs on their list of starches.)
• **What must happen to the bread before your body can use it as fuel?** (The chains that make up the starches must be broken.)
• **What is the purpose of using the iodine?** (The iodine is used as an indicator since one cannot see whether or not starch exists.)

PROCEDURE

3. ■ The strip of bread turns bluish-purple.
4. ■ The center of the bread will turn bluish purple while the toasted outside remains unchanged.

■ The untoasted part of the bread turns bluish-purple, showing that it is mostly starch. The toasted part remains unchanged. This must mean that the toasting process causes a chemical reaction that changes starch into another substance. Students who do a little research will find that the starch in the toasted areas has been changed by heat into dextrin. Dextrin iodide is not bluish-purple.

■ Endothermic because the bread absorbs heat from the toaster to cause starch to change into another substance.

■ Toasting is a step in the digestion of bread. Because toast is partially digested when compared with plain bread, it is actually easier on the digestive process because the first steps have already begun and the body has to do less work in breaking it down.

Activity Bank

TOASTING TO GOOD HEALTH

BEFORE THE ACTIVITY

Gather the materials at least one day prior to the activity. The toasted bread does not have to be warm.

PRE-ACTIVITY DISCUSSION

You may want to begin a brief discussion about digestion and nutrition with stu-
dents. It may be helpful to conclude the discussion with questions similar to the following.
• **What does the digestive system do?** (It breaks down food into simpler substances for use by the body.)
• **Where does digestion begin?** (In the mouth. Teeth are responsible for mechanical digestion, the physical action of breaking down food. Saliva is responsible for chemical digestion.)
• **Do you know what a starch is?** (Starches are a type of carbohydrate. A starch is

Activity Bank

BEFORE THE ACTIVITY

Gather the material at least one day prior to the activity. Make sure you have enough materials for each group in the class.

PRE-ACTIVITY DISCUSSION

Review the concepts of acids and bases. Lead a discussion, perhaps by asking questions similar to the following.

• **What are the properties of an acid?** (Acids have a sour taste, they turn blue litmus paper red, they react with active metals, they contain hydrogen, and they are proton donors.)

• **What are the properties of a base?** (Bases usually taste bitter, are slippery to the touch, turn red litmus paper blue, and are proton acceptors.)

Have students read through the entire activity. You may then want to continue your discussion by asking further questions such as:

• **What are you going to do in this activity?** (Lead students to realize that they are going to test substances to find out whether they are acids or bases using blackberry jam.)

• **What is the purpose of the ammonia, lemon juice, and vinegar?** (They are the substances being tested.)

SAFETY TIPS

Warn students not to inhale the vapors from the ammonia.

TEACHING STRATEGY

1. Have some groups use lemon juice and others use vinegar so that enough students see the results of each.

2. You may wish to have students predict whether the substances are acids or bases. They may already know the answers. Then have them compare their observations with their predictions.

DISCOVERY STRATEGIES

Discuss how the activity relates to the chapter ideas by asking questions similar to the following.

• **What is the purpose of litmus paper?** (Litmus paper is used as an acid–base in-

---

You have probably seen or used litmus paper to determine whether a substance is an acid or a base. But did you know that litmus paper is not the only material that can be used as an acid-base indicator? It may surprise you to learn that many foods can also do the job in a pinch! In this activity you will experiment with just such an indicator.

Materials

blackberry jam (a spoonful is enough)
warm water
small drinking glass
household ammonia (several drops)
lemon juice or vinegar (several drops)
spoon
medicine dropper

Procedure

1. Fill the drinking glass half-full with warm tap water.

2. Put a spoonful of jam into the water and gently stir it with the spoon until it is dissolved. The water-jam solution should turn a reddish color.

3. Use the medicine dropper to put a few drops of ammonia into the solution. Stir the solution once or twice. What happens to the color of the solution?

4. Clean the medicine dropper. Use the clean dropper to add several drops of lemon juice or vinegar to the solution. Clean the spoon and again stir the solution. What happens to the color of the solution this time?

---

5. Compare your observations with those of your classmates who added the substance that you did not—lemon juice or vinegar. What happened to their solutions?

■ The jam solution is red when an acid is added to it and greenish-purple when a base is added. From your experiment determine whether ammonia, vinegar, and lemon juice are acids or bases.

The Next Step

Repeat the experiment several more times, each time using a different test substance. You may choose such substances as milk, juice, soda, or fruit. Be sure to clean the spoon between each stirring. Make a chart showing which substances are acids and which are bases. Combine your observations with those of your classmates.

---

dicator.)

• **What was used in the activity instead of litmus paper?** (The blackberry jam.)

• **From your observations, which substances are proton acceptors? Proton donors?** (Lemon juice and vinegar are proton donors; ammonia is a proton acceptor.)

PROCEDURE

3. The color changes to a greenish purple.

4. The color changes back to red. Either substance will turn the solution red.

■ Ammonia is a base; lemon juice and vinegar are acids.

THE NEXT STEP

Check student charts for accuracy.

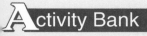

OIL SPILL

You have probably seen television news reports or read newspaper articles about the devastation caused by an oil spill from a supertanker or other holding vessel. But initial reports often underestimate the full spectrum of the damage. In this activity you will simulate interactions with oil so that you can more clearly understand the dangerous consequences of an oil spill.

Materials

medicine dropper
small graduated cylinder
motor oil, used
fan
tongs
3 hard-boiled eggs, not peeled
paper towels
shallow baking pan, about 40 cm × 20 cm
white paper, 1 sheet
graph paper, 1-cm grid
beaker or jar (must be able to hold 3 eggs)

Procedure

1. Partially fill a shallow baking pan two-thirds full with water.

2. Pour the motor oil into the graduated cylinder.

3. Use the medicine dropper to remove 1 mL of oil from the graduated cylinder. Gently squeeze the oil out of the dropper into the center of the pan of water. Describe the interaction between the water and oil.

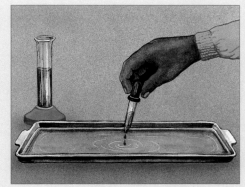

4. Mark off a region on the graph paper that is the same size as the baking pan. After several minutes, sketch the arrangement of oil in the pan of water. When you are finished drawing, count up the number of squares on the graph paper covered by oil. Remember, the area now covered was produced by only 1 mL of oil! Assuming that oil always spreads proportionately, make a chart showing the area that would be covered by 2 mL, 10 mL, 100 mL, and 1 L.

5. Place a fan beside the pan of water and oil. Turn it on and determine if the flow of air affects the spread of oil. What do you discover?

6. Now try shaking the pan slightly. Be careful not to spill the contents. Does this reaction affect the oil at all?

(continued)

Activity Bank

OIL SPILL

BEFORE THE ACTIVITY

Gather the materials at least one day prior to the activity. Make sure you have enough for all the groups in the class. The used motor oil was suggested because it is easily observed and it prevents waste; however if necessary other types of oil could be used. If you prefer, food coloring and water can be used to simulate oil.

PRE-ACTIVITY DISCUSSION

You may want to begin a discussion about oil and oil spills. It may help to ask questions similar to the following.
• **Have you ever seen an oily pan filled with water in your kitchen sink?** (Most students have seen such a pan.)
• **How would you describe the oil and water in the pan?** (The oil would have floated across the top of the water and would have spread out to cover the entire surface of the water.)
• **What happens in salad dressings made up of oil and other liquids, including water?** (The oil does not mix with the other substances, but instead floats at a level depending upon the liquids used.)
• **Are you familiar with any major oil spills?** (Students should remember the Valdez spill in Prince William Sound in Alaska or the deliberate spill during the Gulf War in Kuwait.)
• **What kinds of damage can an oil spill cause?** (Encourage students to describe the terrible destruction caused by an oil spill.)

TEACHING STRATEGY

1. You may wish to set up stations and have groups switch places after each part of the activity in order to cut down on materials.

2. In the first part of the activity, students will learn how a small amount of oil can spread over a huge area. Advise students to draw a graph to match the size of the container before they drop the oil into it. Assist them in extrapolating results for larger volumes of oil, but remind them that these are only crude estimates. Varying situations will make the numbers better or worse.

DISCOVERY STRATEGIES

Discuss how the activity relates to the chapter ideas by asking questions similar to the following.
• **Why is oil so important?** (Oil is used as a source of fuel or lubrication in many processes.)
• **Why does it have to be transported over great distances?** (Oil is found only in certain parts of the world. It must then be transported to the locations where it is used.)
• **Can you suggest advantages and disadvantages to developing a replacement for oil?** (Accept all logical answers.)
• **Can you propose ways of preventing oil spills from supertankers?** (Accept all logical answers.)

PROCEDURE

3. The oil and water do not mix. The oil spreads out across the surface of the water.
5. The moving air makes the oil disperse even further.

6. Shaking the water and oil combination should make the oil spread out further over the water. If the water is already covered, students may wish to begin again.
8. Students should be able to see that oil is beginning to seep through the shell.
9. More and more oil seeps through the longer the eggs remain in the oil. By the last egg, oil has seeped through the shell.

THE BIG PICTURE

1. A small volume of oil, such as 1 mL or 2 mL, may fit in a medicine dropper but when dumped into water, it quickly covers an area many times its original size. This situation is only magnified when the original volume of oil is increased. Thus an oil spill in the ocean cannot be an isolated event. The oil will spread and affect a tremendous area.
2. Blowing air and shaking the water intensified the spread of the oil. These conditions represent the winds that blow ocean waters and the waves that rock the waters. Because wind and waves are natural conditions of oceans, any oil spill will have more far-reaching ramifications than it would under ideal conditions and can be made worse when these conditions are particularly severe.
3. The baby birds forming within the eggs will be altered when the oil seeps into their food supply, which is sealed inside the egg.

7. Gently place the three hard boiled eggs in the jar or beaker. Pour oil into the container until it is full. Place the container under a strong light.

8. After 5 minutes use the tongs to carefully remove one egg. Remove the excess oil with a paper towel. Peel the egg. What do you observe?

9. Remove the second egg after 15 minutes. Peel this egg and record what you observe. Remove the third egg after 30 minutes. Again peel the egg and record your observations.

The Big Picture

1. Supertankers carry millions of liters of oil. In light of your calculations, what can you say about the implications of a large oil spill?

2. What did you learn by blowing air on the oil and by shaking the water? What conditions did these procedures represent? How do these conditions affect the severity of oil spills?

3. What effect could oil have on the eggs of birds nesting near ocean water that becomes contaminated with oil?

THE DOMINO EFFECT

Have you ever played with dominoes? If so, you know that dominoes can be arranged into all sorts of complicated patterns that enable you to knock them all down from a gentle tap on just one domino. Beyond playing, the falling action of dominoes can be used to represent a very complex phenomenon—a nuclear chain reaction. In this activity, you will need 15 dominoes and a stopwatch to learn more about nuclear chemistry.

Procedure

1. Place the dominoes in a row so that each one is standing on its narrow end. Each domino should be about 1–1.5 cm from the next one.
2. Gently tip the first domino in the line over so that it falls on the one behind it. You have just initiated a chain reaction.
   - What keeps the reaction going?
   - How can the row of dominoes be likened to a nuclear chain reaction?
3. Now arrange the dominoes as shown in the accompanying figure.

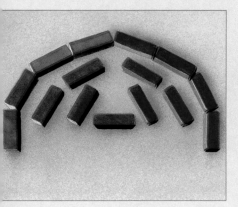

4. Gently tip over the center domino. How does this arrangement differ from the first one?
   - If the dominoes again represent atomic nuclei, how is this chain reaction different from the first one?
5. Which arrangement of dominoes do you think falls faster? Find out. Arrange the dominoes back into a single row. Use the stopwatch to measure the length of time from when you tip the first domino until the last domino falls. Record the measurement. Return the dominoes to the second pattern. Again record the time from the tip of the first domino to the fall of the last one. Which arrangement falls faster?
6. Now set up the dominoes as shown in the accompanying figure.

7. Tip the leftmost domino. Record the amount of time it takes for the dominoes to fall.
   - How does the length of time it takes for the dominoes to fall in this pattern compare with the times for the other two patterns?

(continued)

are uncontrolled while those in a nuclear power plant are carefully controlled. Chain reactions such as those in an atomic bomb could not occur in a nuclear power plant because the same opportunities for continuous fission are not there.)

TEACHING STRATEGY

1. Most students will have had experience setting up dominoes, but some may find it troublesome. Circulate around the classroom to assist students having difficulty setting up the dominoes properly.
2. When timing each arrangement, suggest the class select one student to use the stopwatch and another to knock over the dominoes. A third student can record the measurement. Suggest that they take at least two measurements for each arrangement.

DISCOVERY STRATEGIES

Discuss how the activity relates to the ideas in the chapter by asking questions similar to the following.

- **In light of this activity, if the particles released when a nucleus splits in half are somehow absorbed, how will the chain reaction be affected?** (As was seen when a domino fell and was not able to hit into another one, absorbed particles would be prevented from starting new fission reactions.)
- **What are the advantages of nuclear power? Disadvantages?** (Accept all reasonable answers.)

PROCEDURE

2. ■ Each domino hits the next one until the last domino falls.

■ The dominoes can be used to represent atomic nuclei. When the first domino is tipped, it is like an atomic nucleus being struck by a neutron. The neutron causes the nucleus to split into two smaller nuclei. When this happens at least one neutron is released. The neutron goes on to hit another nucleus. This is what is being represented each time a domino falls and hits the next one. The process continues until the last nucleus is split—or in this case, the last domino falls.

4. In this case, each domino hits two more dominoes when it falls. In turn, the two falling dominoes tip four more over. Each falling domino results in more dominoes falling over in each successive row.

Activity Bank

THE DOMINO EFFECT

BEFORE THE ACTIVITY

Gather the materials at least one day prior to the activity. Make sure you have enough materials for all groups in the class.

PRE-ACTIVITY DISCUSSION

Review the concepts of nuclear fission, chain reactions, and nuclear reactors. You may want to begin a discussion by asking questions similar to the following.

- **What is nuclear fission?** (Remind students that fission is the splitting of an atomic nucleus into two smaller nuclei of approximately equal mass.)
- **What is a nuclear chain reaction?** (A continuous series of fission reactions.)
- **What is the difference between fission reactions in an atomic bomb and those in a nuclear power plant?** (Lead students to realize that the reactions in atomic bombs

■ This reaction represents the situation in which a fissioning atom releases two neutrons each time. Those two neutrons go on to strike two more nuclei.

5. The second pattern. The first arrangement, the single line, takes longer to fall because only one domino falls at a time.

6. ■ The length of time is shorter than that for the straight row to fall but longer than that for the semicircular configuration to fall.

■ Students should realize that in this configuration, eight dominoes fall without hitting other dominoes. Thus the main row of dominoes—those set up in a slanted row—represent nuclei that each release two neutrons. One neutron splits another nuclei that releases two more neutrons. Those two neutrons hit two more nuclei. The other neutron hits a nucleus that releases another neutron. This neutron, however, does not strike another nucleus. This slows down the reaction.

WHAT THIS ALL MEANS

■ Nuclear reactions that occur in nuclear reactors must be carefully controlled. A fast, uncontrolled reaction is not desirable.

■ Nuclear fuel rods (dominoes) can be arranged in such a way that released neutrons are either utilized or wasted. In addition, substances that absorb neutrons can be used to prevent released neutrons from striking additional nuclei. A little research will show that some nuclear power plants submerge nuclear fuel rods in water containing boric acid. Boric acid absorbs neutrons. The rate of nuclear reactions can be controlled by altering the amount of boric acid in the water.

■ How can you explain this arrangement in terms of a nuclear chain reaction?

8. Design your own arrangement for the dominoes. Determine how long it takes for all the dominoes to fall from this arrangement. Compare this time with those of the other arrangements and with those of your classmates. Make a chart or a poster showing the different arrangements and the length of time recorded for each one.

What This All Means

■ Why might it be important to slow down a reaction?

■ After completing this activity, can you think of how nuclear chain reactions can be controlled in nuclear power plants?

Appendix A

The metric system of measurement is used by scientists throughout the world. It is based on units of ten. Each unit is ten times larger or ten times smaller than the next unit. The most commonly used units of the metric system are given below. After you have finished reading about the metric system, try to put it to use. How tall are you in metrics? What is your mass? What is your normal body temperature in degrees Celsius?

Commonly Used Metric Units

Length   The distance from one point to another

meter (m)   A meter is slightly longer than a yard.
              1 meter = 1000 millimeters (mm)
              1 meter = 100 centimeters (cm)
              1000 meters = 1 kilometer (km)

Volume   The amount of space an object takes up

liter (L)   A liter is slightly more than a quart.
              1 liter = 1000 milliliters (mL)

Mass   The amount of matter in an object

gram (g)   A gram has a mass equal to about one paper clip.

              1000 grams = 1 kilogram (kg)

Temperature   The measure of hotness or coldness

degrees   0°C = freezing point of water
Celsius (°C)   100°C = boiling point of water

Metric–English Equivalents

2.54 centimeters (cm) = 1 inch (in.)
1 meter (m) = 39.37 inches (in.)
1 kilometer (km) = 0.62 miles (mi)
1 liter (L) = 1.06 quarts (qt)
250 milliliters (mL) = 1 cup (c)
1 kilogram (kg) = 2.2 pounds (lb)
28.3 grams (g) = 1 ounce (oz)
°C = 5/9 × (°F – 32)

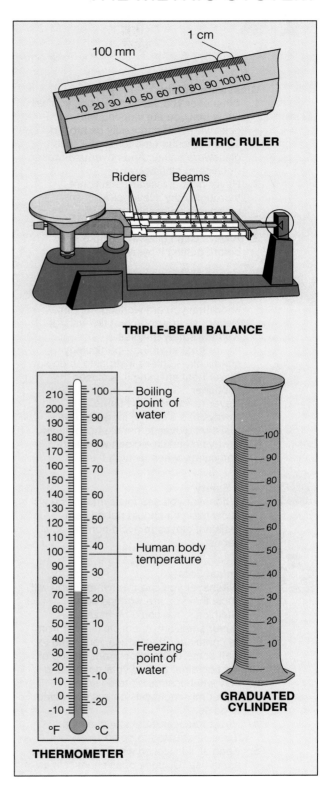

METRIC RULER

Riders   Beams

TRIPLE-BEAM BALANCE

Boiling point of water

Human body temperature

Freezing point of water

°F   °C

THERMOMETER

GRADUATED CYLINDER

Appendix B

Glassware Safety

1. Whenever you see this symbol, you will know that you are working with glassware that can easily be broken. Take particular care to handle such glassware safely. And never use broken or chipped glassware.
2. Never heat glassware that is not thoroughly dry. Never pick up any glassware unless you are sure it is not hot. If it is hot, use heat-resistant gloves.
3. Always clean glassware thoroughly before putting it away.

Fire Safety

1. Whenever you see this symbol, you will know that you are working with fire. Never use any source of fire without wearing safety goggles.
2. Never heat anything—particularly chemicals—unless instructed to do so.
3. Never heat anything in a closed container.
4. Never reach across a flame.
5. Always use a clamp, tongs, or heat-resistant gloves to handle hot objects.
6. Always maintain a clean work area, particularly when using a flame.

Heat Safety

Whenever you see this symbol, you will know that you should put on heat-resistant gloves to avoid burning your hands.

Chemical Safety

1. Whenever you see this symbol, you will know that you are working with chemicals that could be hazardous.
2. Never smell any chemical directly from its container. Always use your hand to waft some of the odors from the top of the container toward your nose—and only when instructed to do so.
3. Never mix chemicals unless instructed to do so.
4. Never touch or taste any chemical unless instructed to do so.
5. Keep all lids closed when chemicals are not in use. Dispose of all chemicals as instructed by your teacher.

6. Immediately rinse with water any chemicals, particularly acids, that get on your skin and clothes. Then notify your teacher.

Eye and Face Safety

1. Whenever you see this symbol, you will know that you are performing an experiment in which you must take precautions to protect your eyes and face by wearing safety goggles.
2. When you are heating a test tube or bottle, always point it away from you and others. Chemicals can splash or boil out of a heated test tube.

Sharp Instrument Safety

1. Whenever you see this symbol, you will know that you are working with a sharp instrument.
2. Always use single-edged razors; double-edged razors are too dangerous.
3. Handle any sharp instrument with extreme care. Never cut any material toward you; always cut away from you.
4. Immediately notify your teacher if your skin is cut.

Electrical Safety

1. Whenever you see this symbol, you will know that you are using electricity in the laboratory.
2. Never use long extension cords to plug in any electrical device. Do not plug too many appliances into one socket or you may overload the socket and cause a fire.
3. Never touch an electrical appliance or outlet with wet hands.

Animal Safety

1. Whenever you see this symbol, you will know that you are working with live animals.
2. Do not cause pain, discomfort, or injury to an animal.
3. Follow your teacher's directions when handling animals. Wash your hands thoroughly after handling animals or their cages.

Appendix C

Appendix C

SCIENCE SAFETY RULES

One of the first things a scientist learns is that working in the laboratory can be an exciting experience. But the laboratory can also be quite dangerous if proper safety rules are not followed at all times. To prepare yourself for a safe year in the laboratory, read over the following safety rules. Then read them a second time. Make sure you understand each rule. If you do not, ask your teacher to explain any rules you are unsure of.

Dress Code

1. Many materials in the laboratory can cause eye injury. To protect yourself from possible injury, wear safety goggles whenever you are working with chemicals, burners, or any substance that might get into your eyes. Never wear contact lenses in the laboratory.

2. Wear a laboratory apron or coat whenever you are working with chemicals or heated substances.

3. Tie back long hair to keep it away from any chemicals, burners and candles, or other laboratory equipment.

4. Remove or tie back any article of clothing or jewelry that can hang down and touch chemicals and flames.

General Safety Rules

5. Read all directions for an experiment several times. Follow the directions exactly as they are written. If you are in doubt about any part of the experiment, ask your teacher for assistance.

6. Never perform activities that are not authorized by your teacher. Obtain permission before "experimenting" on your own.

7. Never handle any equipment unless you have specific permission.

8. Take extreme care not to spill any material in the laboratory. If a spill occurs, immediately ask your teacher about the proper cleanup procedure. Never simply pour chemicals or other substances into the sink or trash container.

9. Never eat in the laboratory.

10. Wash your hands before and after each experiment.

First Aid

11. Immediately report all accidents, no matter how minor, to your teacher.

12. Learn what to do in case of specific accidents, such as getting acid in your eyes or on your skin. (Rinse acids from your body with lots of water.)

13. Become aware of the location of the first-aid kit. But your teacher should administer any required first aid due to injury. Or your teacher may send you to the school nurse or call a physician.

14. Know where and how to report an accident or fire. Find out the location of the fire extinguisher, phone, and fire alarm. Keep a list of important phone numbers—such as the fire department and the school nurse—near the phone. Immediately report any fires to your teacher.

Heating and Fire Safety

15. Again, never use a heat source, such as a candle or burner, without wearing safety goggles.

16. Never heat a chemical you are not instructed to heat. A chemical that is harmless when cool may be dangerous when heated.

17. Maintain a clean work area and keep all materials away from flames.

18. Never reach across a flame.

19. Make sure you know how to light a Bunsen burner. (Your teacher will demonstrate the proper procedure for lighting a burner.) If the flame leaps out of a burner toward you, immediately turn off the gas. Do not touch the burner. It may be hot. And never leave a lighted burner unattended!

20. When heating a test tube or bottle, always point it away from you and others. Chemicals can splash or boil out of a heated test tube.

21. Never heat a liquid in a closed container. The expanding gases produced may blow the container apart, injuring you or others.

O ■ 163

22. Before picking up a container that has been heated, first hold the back of your hand near it. If you can feel the heat on the back of your hand, the container may be too hot to handle. Use a clamp or tongs when handling hot containers.

Using Chemicals Safely

23. Never mix chemicals for the "fun of it." You might produce a dangerous, possibly explosive substance.

24. Never touch, taste, or smell a chemical unless you are instructed by your teacher to do so. Many chemicals are poisonous. If you are instructed to note the fumes in an experiment, gently wave your hand over the opening of a container and direct the fumes toward your nose. Do not inhale the fumes directly from the container.

25. Use only those chemicals needed in the activity. Keep all lids closed when a chemical is not being used. Notify your teacher whenever chemicals are spilled.

26. Dispose of all chemicals as instructed by your teacher. To avoid contamination, never return chemicals to their original containers.

27. Be extra careful when working with acids or bases. Pour such chemicals over the sink, not over your workbench.

28. When diluting an acid, pour the acid into water. Never pour water into an acid.

29. Immediately rinse with water any acids that get on your skin or clothing. Then notify your teacher of any acid spill.

Using Glassware Safely

30. Never force glass tubing into a rubber stopper. A turning motion and lubricant will be helpful when inserting glass tubing into rubber stoppers or rubber tubing. Your teacher will demonstrate the proper way to insert glass tubing.

31. Never heat glassware that is not thoroughly dry. Use a wire screen to protect glassware from any flame.

32. Keep in mind that hot glassware will not appear hot. Never pick up glassware without first checking to see if it is hot. See #22.

33. If you are instructed to cut glass tubing, fire-polish the ends immediately to remove sharp edges.

34. Never use broken or chipped glassware. If glassware breaks, notify your teacher and dispose of the glassware in the proper trash container.

35. Never eat or drink from laboratory glassware. Thoroughly clean glassware before putting it away.

Using Sharp Instruments

36. Handle scalpels or razor blades with extreme care. Never cut material toward you; cut away from you.

37. Immediately notify your teacher if you cut your skin when working in the laboratory.

Animal Safety

38. No experiments that will cause pain, discomfort, or harm to mammals, birds, reptiles, fishes, and amphibians should be done in the classroom or at home.

39. Animals should be handled only if necessary. If an animal is excited or frightened, pregnant, feeding, or with its young, special handling is required.

40. Your teacher will instruct you as to how to handle each animal species that may be brought into the classroom.

41. Clean your hands thoroughly after handling animals or the cage containing animals.

End-of-Experiment Rules

42. After an experiment has been completed, clean up your work area and return all equipment to its proper place.

43. Wash your hands after every experiment.

44. Turn off all burners before leaving the laboratory. Check that the gas line leading to the burner is off as well.

THE CHEMICAL ELEMENTS

NAME	SYMBOL	ATOMIC NUMBER	ATOMIC MASS†	NAME	SYMBOL	ATOMIC NUMBER	ATOMIC MASS†
Actinium	Ac	89	(227)	Neodymium	Nd	60	144.2
Aluminum	Al	13	27.0	Neon	Ne	10	20.2
Americium	Am	95	(243)	Neptunium	Np	93	(237)
Antimony	Sb	51	121.8	Nickel	Ni	28	58.7
Argon	Ar	18	39.9	Niobium	Nb	41	92.9
Arsenic	As	33	74.9	Nitrogen	N	7	14.01
Astatine	At	85	(210)	Nobelium	No	102	(255)
Barium	Ba	56	137.3	Osmium	Os	76	190.2
Berkelium	Bk	97	(247)	Oxygen	O	8	16.00
Beryllium	Be	4	9.01	Palladium	Pd	46	106.4
Bismuth	Bi	83	209.0	Phosphorus	P	15	31.0
Boron	B	5	10.8	Platinum	Pt	78	195.1
Bromine	Br	35	79.9	Plutonium	Pu	94	(244)
Cadmium	Cd	48	112.4	Polonium	Po	84	(210)
Calcium	Ca	20	40.1	Potassium	K	19	39.1
Californium	Cf	98	(251)	Praseodymium	Pr	59	140.9
Carbon	C	6	12.01	Promethium	Pm	61	(145)
Cerium	Ce	58	140.1	Protactinium	Pa	91	(231)
Cesium	Cs	55	132.9	Radium	Ra	88	(226)
Chlorine	Cl	17	35.5	Radon	Rn	86	(222)
Chromium	Cr	24	52.0	Rhenium	Re	75	186.2
Cobalt	Co	27	58.9	Rhodium	Rh	45	102.9
Copper	Cu	29	63.5	Rubidium	Rb	37	85.5
Curium	Cm	96	(247)	Ruthenium	Ru	44	101.1
Dysprosium	Dy	66	162.5	Samarium	Sm	62	150.4
Einsteinium	Es	99	(254)	Scandium	Sc	21	45.0
Erbium	Er	68	167.3	Selenium	Se	34	79.0
Europium	Eu	63	152.0	Silicon	Si	14	28.1
Fermium	Fm	100	(257)	Silver	Ag	47	107.9
Fluorine	F	9	19.0	Sodium	Na	11	23.0
Francium	Fr	87	(223)	Strontium	Sr	38	87.6
Gadolinium	Gd	64	157.2	Sulfur	S	16	32.1
Gallium	Ga	31	69.7	Tantalum	Ta	73	180.9
Germanium	Ge	32	72.6	Technetium	Tc	43	(97)
Gold	Au	79	197.0	Tellurium	Te	52	127.6
Hafnium	Hf	72	178.5	Terbium	Tb	65	158.9
Helium	He	2	4.00	Thallium	Tl	81	204.4
Holmium	Ho	67	164.9	Thorium	Th	90	232.0
Hydrogen	H	1	1.008	Thulium	Tm	69	168.9
Indium	In	49	114.8	Tin	Sn	50	118.7
Iodine	I	53	126.9	Titanium	Ti	22	47.9
Iridium	Ir	77	192.2	Tungsten	W	74	183.9
Iron	Fe	26	55.8	Unnilennium	Une	109	(266?)
Krypton	Kr	36	83.8	Unnilhexium	Unh	106	(263)
Lanthanum	La	57	138.9	Unniloctium	Uno	108	(265)
Lawrencium	Lr	103	(256)	Unnilpentium	Unp	105	(262)
Lead	Pb	82	207.2	Unnilquadium	Unq	104	(261)
Lithium	Li	3	6.94	Unnilseptium	Uns	107	(262)
Lutetium	Lu	71	175.0	Uranium	U	92	238.0
Magnesium	Mg	12	24.3	Vanadium	V	23	50.9
Manganese	Mn	25	54.9	Xenon	Xe	54	131.3
Mendelevium	Md	101	(258)	Ytterbium	Yb	70	173.0
Mercury	Hg	80	200.6	Yttrium	Y	39	88.9
Molybdenum	Mo	42	95.9	Zinc	Zn	30	65.4
				Zirconium	Zr	40	91.2

†Numbers in parentheses give the mass number of the most stable isotope.

1

1	
H	
Hydrogen	
1.00794	

Key

6	Atomic number
C	Element's symbol
Carbon	Element's name
12.011	Atomic mass

2

1		
1		
3	4	
Li	**Be**	
Lithium	Beryllium	
6.941	9.0122	

Transition Metals

3

2			3	4	5	6	7	8	9
11	12								
Na	**Mg**								
Sodium	Magnesium								
22.990	24.305								

4

19	20	21	22	23	24	25	26	27
K	**Ca**	**Sc**	**Ti**	**V**	**Cr**	**Mn**	**Fe**	**Co**
Potassium	Calcium	Scandium	Titanium	Vanadium	Chromium	Manganese	Iron	Cobalt
39.098	40.08	44.956	47.88	50.94	51.996	54.938	55.847	58.9332

5

37	38	39	40	41	42	43	44	45
Rb	**Sr**	**Y**	**Zr**	**Nb**	**Mo**	**Tc**	**Ru**	**Rh**
Rubidium	Strontium	Yttrium	Zirconium	Niobium	Molybdenum	Technetium	Ruthenium	Rhodium
85.468	87.62	88.9059	91.224	92.91	95.94	(98)	101.07	102.906

6

55	56		72	73	74	75	76	77
Cs	**Ba**	57 to 71	**Hf**	**Ta**	**W**	**Re**	**Os**	**Ir**
Cesium	Barium		Hafnium	Tantalum	Tungsten	Rhenium	Osmium	Iridium
132.91	137.33		178.49	180.95	183.85	186.207	190.2	192.22

7

87	88		104	105	106	107	108	109
Fr	**Ra**	89 to 103	**Unq**	**Unp**	**Unh**	**Uns**	**Uno**	**Une**
Francium	Radium		Unnilquadium	Unnilpentium	Unnilhexium	Unnilseptium	Unniloctium	Unnilennium
(223)	226.025		(261)	(262)	(263)	(262)	(265)	(266)

Rare-Earth Elements

Lanthanoid Series

57	58	59	60	61	62
La	**Ce**	**Pr**	**Nd**	**Pm**	**Sm**
Lanthanum	Cerium	Praseodymium	Neodymium	Promethium	Samarium
138.906	140.12	140.908	144.24	(145)	150.36

Actinoid Series

89	90	91	92	93	94
Ac	**Th**	**Pa**	**U**	**Np**	**Pu**
Actinium	Thorium	Protactinium	Uranium	Neptunium	Plutonium
227.028	232.038	231.036	238.029	237.048	(244)

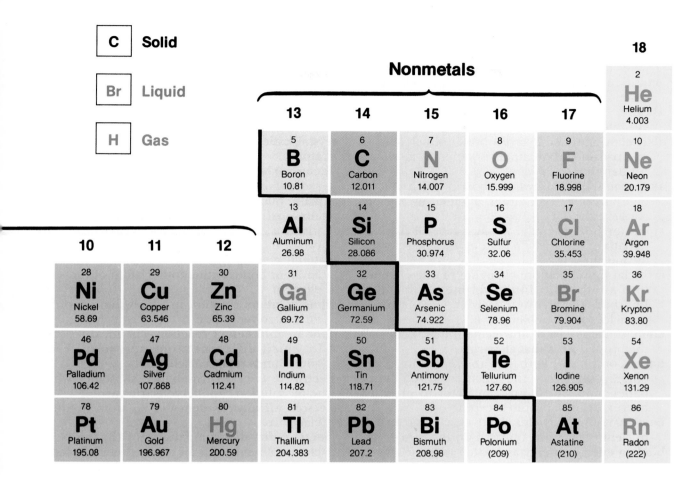

			Nonmetals				**18**
							2 **He** Helium 4.003

C	**Solid**
Br	Liquid
H	Gas

13	**14**	**15**	**16**	**17**
5 **B** Boron 10.81	6 **C** Carbon 12.011	7 **N** Nitrogen 14.007	8 **O** Oxygen 15.999	9 **F** Fluorine 18.998	10 **Ne** Neon 20.179

| | | | **13**
 Al
 Aluminum
 26.98 | **14**
 Si
 Silicon
 28.086 | 15
 P
 Phosphorus
 30.974 | 16
 S
 Sulfur
 32.06 | 17
 Cl
 Chlorine
 35.453 | 18
 Ar
 Argon
 39.948 |

10	**11**	**12**
28 **Ni** Nickel 58.69	29 **Cu** Copper 63.546	30 **Zn** Zinc 65.39	31 **Ga** Gallium 69.72	32 **Ge** Germanium 72.59	33 **As** Arsenic 74.922	34 **Se** Selenium 78.96	35 **Br** Bromine 79.904	36 **Kr** Krypton 83.80
46 **Pd** Palladium 106.42	47 **Ag** Silver 107.868	48 **Cd** Cadmium 112.41	49 **In** Indium 114.82	50 **Sn** Tin 118.71	51 **Sb** Antimony 121.75	52 **Te** Tellurium 127.60	53 **I** Iodine 126.905	54 **Xe** Xenon 131.29
78 **Pt** Platinum 195.08	79 **Au** Gold 196.967	80 **Hg** Mercury 200.59	81 **Tl** Thallium 204.383	82 **Pb** Lead 207.2	83 **Bi** Bismuth 208.98	84 **Po** Polonium (209)	85 **At** Astatine (210)	86 **Rn** Radon (222)

The symbols shown here for elements 104-109 are being used temporarily until names for these elements can be agreed upon.

Metals

Mass numbers in parentheses are those of the most stable or common isotope.

63 **Eu** Europium 151.96	64 **Gd** Gadolinium 157.25	65 **Tb** Terbium 158.925	66 **Dy** Dysprosium 162.50	67 **Ho** Holmium 164.93	68 **Er** Erbium 167.26	69 **Tm** Thulium 168.934	70 **Yb** Ytterbium 173.04	71 **Lu** Lutetium 174.967
95 **Am** Americium (243)	96 **Cm** Curium (247)	97 **Bk** Berkelium (247)	98 **Cf** Californium (251)	99 **Es** Einsteinium (252)	100 **Fm** Fermium (257)	101 **Md** Mendelevium (258)	102 **No** Nobelium (259)	103 **Lr** Lawrencium (260)

Glossary

acid: compound with a pH below 7 that tastes sour, turns blue litmus paper red, reacts with metals to produce hydrogen gas, and ionizes in water to produce hydrogen ions; proton donor

activation energy: energy required for a chemical reaction to occur

alkane: straight-chain or branched-chain saturated hydrocarbon

alkene: unsaturated hydrocarbon in which at least one pair of carbon atoms is joined by a double covalent bond

alkyne: unsaturated hydrocarbon in which at least one pair of carbon atoms is joined by a triple covalent bond

alpha (AL-fuh) **particle:** weakest type of nuclear radiation; consists of a helium nucleus released during alpha decay

artificial transmutation: changing of one element into another by unnatural means; involves bombarding a nucleus with high-energy particles to cause change

atom: smallest part of any element that retains all the properties of that element

base: compound with pH above 7 that tastes bitter, is slippery to the touch, turns red litmus paper blue, and ionizes in water to produce hydroxide ions; proton acceptor

beta (BAYT-uh) **particle:** electron, created in the nucleus of an atom, released during beta decay

binding energy: energy associated with the strong nuclear force that holds an atomic nucleus together; related to the stability of a nucleus

bubble chamber: device that uses a superheated liquid to create bubbles when radioactive particles pass through it

catalyst: substance that increases the rate of a chemical reaction without being changed by the reaction

chemical bonding: combining of atoms of elements to form new substances

chemical equation: description of a chemical reaction using symbols to represent elements and formulas to represent equations

chemical reaction: process in which substances undergo physical and chemical changes that result in the formation of new substances with different properties

cloud chamber: device to study radioactivity, which uses a cooled gas that will condense around radioactive particles

collision theory: theory that relates collisions among particles to reaction rate; reaction rate depends on such factors as concentration, surface area, temperature, and catalysts

concentrated solution: solution in which a large amount of solute is dissolved in a solvent

concentration: amount of a solute dissolved in a certain amount of solvent

covalent bonding: bonding that involves the sharing of electrons

crystal lattice: regular, repeating arrangement of atoms

decay series: sequence of steps by which a radioactive nucleus decays into a nonradioactive nucleus

decomposition reaction: chemical reaction in which a complex substance breaks down into two or more simpler substances

diatomic element: element whose atoms can form covalent bonds with another atom of the same element

dilute solution: solution in which there is only a little dissolved solute

double-replacement reaction: chemical reaction in which different atoms in two different compounds replace each other

electrolyte (ee-LEHK-troh-light): substance whose water solution conducts electric current

electron affinity: tendency of an atom to attract electrons

electron-dot diagram: diagram that uses the chemical symbol for an element surrounded by a series of dots to represent the electron sharing that takes place in a covalent bond

electroscope: device consisting of a metal rod with two thin metal leaves at one end that can be used to detect radioactivity

endothermic reaction: chemical reaction in which energy is absorbed

exothermic (ek-soh-THER-mihk) **reaction:** chemical reaction in which energy is released

fraction: petroleum part with its own boiling point

gamma (GAM-uh) **ray:** high-frequency electro-magnetic wave released during gamma decay; strongest type of nuclear radiation

Geiger counter: device that can be used to detect radioactivity because it produces an electric current in the presence of a radioactive substance

half-life: amount of time it takes for half the atoms in a given sample of an element to decay

hydrocarbon: organic compound that contains only hydrogen and carbon

ion: an atom that has become charged due to the loss or gain of electrons

ionic bonding: bonding that involves the transfer of electrons

ionization: process of removing electrons and forming ions

ionization energy: energy required for ionization

isomer: one of a number of compounds that have the same molecular formula but different structures

isotope (IGH-suh-tohp): atom that has the same number of protons (atomic number) as another atom but a different number of neutrons

kinetics: study of the rates of chemical reactions

metallic bond: bond formed by atoms of metals, in which the outer electrons of the atoms form a common electron cloud

molecule (MAHL-ih-kyool): combination of atoms formed by a covalent bond

monomer: smaller molecule that joins with other smaller molecules to form a chain molecule called a polymer

natural polymer: polymer molecule found in nature; for example, cotton, silk, and wool

network solid: covalent substance whose molecules are very large because the atoms involved continue to bond to one another; have rather high melting points

neutralization (noo-truhl-ih-ZAY-shun): reaction in which an acid combines with a base to form a salt and water

nonelectrolyte: substance whose water solution does not conduct electric current

nuclear chain reaction: series of fission reactions that occur because the products released during one fission reaction cause fission reactions in other atoms

nuclear fission (FIHSH-uhn): splitting of an atomic nucleus into two smaller nuclei of approximately equal mass

nuclear fusion (FYOO-zhuhn): joining of two atomic nuclei of smaller mass to form a single nucleus of larger mass

nuclear radiation: particles and energy released from a radioactive nucleus

nuclear strong force: force that overcomes the electric force of repulsion among protons in an atomic nucleus and binds the nucleus together

organic compound: compound that contains carbon

oxidation number: number of electrons an atom gains, loses, or shares when it forms chemical bonds

petrochemical product: product made either directly or indirectly from petroleum

petroleum: substance believed to have been formed hundreds of millions of years ago when dead plants and animals were buried beneath sediments such as mud, sand, silt, or clay at the bottom of the oceans; crude oil

pH: measure of the hydronium ion concentration of a solution; measured on a scale from 0 to 14

polyatomic ion: group of covalently bonded atoms that acts like a single atom when combining with other atoms

polymer: large molecule in the form of a chain whose links are smaller molecules called monomers

polymerization (poh-lihm-er-uh-ZAY-shuhn): process of chemically bonding monomers to form polymers

product: substance produced by a chemical reaction

radioactive: description for a nucleus that gives off nuclear radiation in the form of mass and energy in order to become stable

radioactive decay: process in which a nucleus spontaneously emits particles or rays to become lighter and more stable

radioactivity: release of energy and matter that results from changes in the nucleus of an atom

radioisotope: artificially produced radioactive isotope; often used in medicine or industry

reactant (ree-AK-tuhnt): substance that enters into a chemical reaction

reaction rate: measure of how quickly reactants change into products

refining: process of separating petroleum into its fractions

salt: compound formed from the positive ion of a base and the negative ion of an acid

saturated hydrocarbon: hydrocarbon in which all the bonds between carbon atoms are single covalent bonds

saturated solution: solution that contains all the solute it can hold at a given temperature

single-replacement reaction: chemical reaction in which an uncombined element replaces an element that is part of a compound

solubility (sahl-yoo-BIHL-uh-tee): measure of how much of a solute can be dissolved in a given amount of solvent under certain conditions

solute (SAHL-yoot): substance that is dissolved in a solution

solution: mixture in which one substance is dissolved, or broken down, in another substance

solvent (SAHL-vuhnt): substance in a solution that does the dissolving

structural formula: description of a molecule that shows the kind, number, and arrangement of atoms in a molecule

substituted hydrocarbon: hydrocarbon formed when one or more hydrogen atoms in a hydrocarbon ring or chain is replaced by a different atom or group of atoms

supersaturated solution: unstable solution that holds more solute than is normal for a given temperature

synthesis (SIHN-thuh-sihs) **reaction:** chemical reaction in which two or more simple substances combine to form a new, more complex substance

synthetic polymer: polymer that does not occur naturally; formed from petrochemicals by people

tracer: radioactive element whose pathway can be followed through the steps of a chemical reaction or industrial process

transmutation: process in which one element is changed into another as a result of changes in the nucleus

transuranium element: element formed synthetically; has more than 92 protons in its nucleus

unsaturated hydrocarbon: hydrocarbon in which one or more of the bonds between carbon atoms is a double covalent or triple covalent bond

unsaturated solution: solution that contains less solute than it can possibly hold at a given temperature

valence electron: electron in the outermost energy level of an atom

Index

Credits